cleanair.ca

a citizen's action guide

cleanair.ca

a citizen's action guide

Chris Tollefson

with Chris Rhone and Chris Rolfe

A project of the Environmental Law Centre, University of Victoria

Published by Sierra Legal Defence Fund

Printed in Canada on 60 lb Windsor recycled paper (30% post-consumer waste), with vegetable-based inks.

Canadian Cataloguing in Publication Data

Tollefson, Chris, 1959 –

cleanair.ca : a citizen's action guide

Includes bibliographical references.

ISBN 0-9698351-4-0

1. Air–Pollution. 2. Air–Pollution–Canada. 3. Air–Pollution–Law and legislation–Canada. 4. Air quality management–Canada–Citizen participation. I. Rhone, Christopher, 1971– II. Rolfe, Christopher, 1961– II. Sierra Legal Defence Fund II. University of Victoria (B.C.). Environmental Law Centre III. Title. IV. Title: cleanair.ca.

TD883.7.C3T65 2000 363.739,2,0971 C00-900960-4

Written by Chris Tollefson, Chris Rhone, and Chris Rolfe
Designed by Working Design
Photo on page 22 by Kelly Reinhardt

The Environmental Law Centre gratefully acknowledges the Law Foundation of British Columbia, West Coast Environmental Law Association, the University of Victoria Eco-Research Chair in Environmental Law and Policy, the University of Victoria Students' Society, Environment Canada, and the Sierra Legal Defence Fund for their support of this project.

The Sierra Legal Defence Fund would like to thank the EJLB Foundation for its generous support of SLDF's work on air pollution.

Environmental Law Centre
University of Victoria
Faculty of Law
McGill Road
Victoria, BC V8W 3H7
250.721.8188
Fax: 250.472.4528
elc@uvic.ca
www.cleanair.ca

contents

chapter 4 **motor vehicle air pollution**

How we can reduce our dependence on the motor vehicle, and reduce the level of harmful emissions per motor vehicle kilometre travelled.

chapter 5 **industrial sources of air pollution**

How governments currently regulate and manage industrial source air pollution, and how we can collectively – through enhanced citizen participation and other means – do a better job combatting it.

chapter 6 **a citizen's toolkit**

The tools you need to get started, and be effective, as an advocate for clean air.

Appendix

A comprehensive reference guide including an extensive appendix of online information sources, and ways to monitor and contact government agencies and decisionmakers.

preface

Never before have Canadians been as concerned about clean air. The federal government estimates that air pollution causes 5,000 premature deaths in Canada each year. Environmental and public health advocates believe the number is closer to 16,000. If they are right, air pollution is responsible for approximately forty premature deaths across Canada each and every day.

The magnitude and complexity of the threats to our clean air are staggering. Some of these threats – including smog, acid rain, and hazardous air pollutants – tend to be local or regional in scope. Others – such as ozone layer depletion and climate change – are truly global in nature.

The need for action at all levels and in all sectors of society, within and beyond our borders, is undeniable. If solutions are to be found, citizens everywhere must become more effective advocates for clean air.

This guide has been written in the belief that citizens can make a difference. Making a difference, however, depends heavily on being informed and strategic in advocating for change. This guide is intended to provide an accessible way for citizens across Canada to become informed about air issues and to serve as a practical reference source for those working to protect our air resources.

The science, law, and policy surrounding air issues is not only dauntingly complex but also rapidly changing. To stay abreast of these developments, we have identified and compiled a list of key online sources. In addition, as the title of the guide suggests, the Environmental Law Centre at the University of Victoria, together with several partner organizations across the country, will be launching a citizen's clean air website: www.cleanair.ca.

We hope that this site will provide a place for groups and individuals concerned about air pollution to discuss clean air issues, share information and strategies, and coordinate efforts to better protect our air. We invite you to visit us there soon.

Chris Tollefson
Executive Director, Environmental Law Centre
Faculty of Law, University of Victoria

acknowledgments

When the Environmental Law Centre opened its doors in 1995, becoming Canada's first public interest environmental law clinic, a key priority was to assist citizens and community groups working on important environmental issues that had yet to capture significant public attention. One of the first, and certainly our most long-standing, priorities was to publish an accessible, action-oriented primer on the law and science of clean air.

Initial research was carried out as a collaborative clinic project in 1997 during the first year the ELC was offered as a course at the University of Victoria, Faculty of Law. As this work was reaching completion, having secured seed funding from the Law Foundation of British Columbia and being encouraged by the response of citizens groups from across the country, we decided to produce a more comprehensive guide aimed at a national audience.

During this second stage, the project received funding from the UVic Eco-Research Chair in Environmental Law and Policy and the UVic Students' Society, as well as considerable financial and technical support from the West Coast Environmental Law Association. As a new and expanded manuscript took shape, the Sierra Legal Defence Fund agreed to publish and distribute the guide, the aims of which dovetailed closely with their expanding docket of litigation nationwide on air issues.

At this juncture, the project expanded yet again. To ensure that the guide reached its intended audience and to maximize opportunities for groups to share information and stay up to date in this rapidly evolving area, we decided to develop an interactive national website on clean air issues. Support to design this website, including resources to conduct a national survey of groups active around air quality issues, was provided by Environment Canada. The website – www.cleanair.ca – is being launched concurrently with publication of this book.

Through it all, we have received support, encouragement, and assistance from many sources. Recognition is owed first to the many ELC students past and present who have toiled at various points on the research and funding work that underpins this project: Bradley Bryan, Harry Djorgee, Michelle Ellison, Brendan Gaunt, Jeffrey Johnston, Karen Levin, Anita McPhee, Maria

Mohammed, Amandeep Singh, Kathy Stothard, Peter Trotzki, and Aaron Welch. In addition, Deborah Curran (formerly of the UVic Eco-Research Chair) deserves particular thanks for her contribution to what is now chapter 3.

We have also benefited from the ongoing advice and feedback of a number of air quality experts. We would especially like to thank Jack Gibbons, Tim Howard, Gerry Scott, Melanie Steiner, John Swaigen, and Mark Winfield, as well as Dr. David Suzuki and Dr. David Bates. In recognizing their respective contributions, we exempt them from responsibility for any errors or omissions in what follows.

Special thanks are also owed to Karen Wristen of SLDF (an unflagging supporter of the project from the very beginning); Mary Beth Berube and Tonya Wilts of Environment Canada; Shelley Kostiuk (our unflappable ELC staff person); and particularly my research assistant Zoran Bozic, who, among other things, deserves credit for the comprehensive appendix at the end of the book.

In the final stages of the project, I had the pleasure of working closely with a number of talented production people, including Randy Schmidt and Cathy Wilkinson (editors); Antonia Banyard and Kris Klaasen of Working Design (layout and graphic design); Jim Boothroyd of SLDF (guide production and launch); Amanda Gibbs of IMPACS (website survey and outreach); and Deryk Wenaus (website development).

This guide would not have been completed without the substantial contributions made by my co-authors: Chris Rhone and Chris Rolfe. This was a harmonious and productive collaboration, due potentially, a sceptic might suggest, to the fact that much of the work was completed independently. Over the last three years, our biggest ongoing challenge has been to surmount the confusion created by our respective parents' preferences in boys' names.

That said, I would like, in closing, to recognize and acknowledge my parents: my father, for what he taught me about the possibilities of law, and my mother for what she taught me about the possibilities of activism.

Chris Tollefson
Victoria, BC
June 26, 2000

cleanair.ca

a citizen's action guide

introduction

chapter 1

Human societies have always had the capacity to affect the world around them. However, it is only in the relatively recent past that we have been able to cause significant harm to our environment, particularly the atmosphere.

Until quite recently, Canadians tended to take clean air for granted. Indeed, for many of us, clean air was something that set Canada, especially urban Canada, apart from the rest of an increasingly polluted world. But times have changed. Today, few Canadians are complacent about air quality. Air pollution has become a key environmental and health priority.

There are many reasons why air pollution has become such a pressing public concern. One is its universality. Air pollution affects us all: our ability to enjoy life and our surroundings, our natural environment, and ultimately our health.[1] At the same time, the impacts of air pollution are unequally shared. Those who are least able to protect themselves – children, the aged, and the sick – are often the most susceptible to being harmed by poor air quality. Finally, air pollution is a problem – both in terms of its causes and potential solutions – that is directly connected to choices we make in our daily lives.

Air pollution does not respect political boundaries. Over 50% of smog in Toronto is blown into the city from sources in the eastern United States. Pesticides used in South America affect the health of Canadians in the far north, due to the long-range atmospheric transport of these chemicals.

Although, in recent years, we have become more aware of air pollution, it is by no means a new issue. In 1948, twenty residents of Donora, Pennsylvania, died when the local steel plant released a lethal emission of sulphur dioxide, carbon monoxide, and metal dust. Nearly 7,000 other residents, half the town's population, became severely ill. In 1952, 4,000 residents of London, England, died when a combination of pollutants and weather conditions combined to produce a "killer fog."[2] And, in 1984, a poisonous gas leak from a Union Carbide factory in Bhopal, India, caused 3,400 deaths and seriously injured several thousands more.

Although there are many opportunities for Canadians to get involved in air issues, there are also many barriers. A key barrier is getting and staying informed. Both the science and law surrounding air pollution is notoriously complex and rapidly evolving. A key goal of this book is to serve as a scientific and legal primer on air pollution, so that concerned citizens can play a more effective role in protecting our air resources. The book is also designed for those who are already informed and involved around air issues. For air quality activists, the book provides current information and tools on air pollution problems and trends, and ways to stay abreast of developments in this ever changing area.

A Short History of Air Pollution

Human societies have always had the capacity to affect the world around them. However, it is only in the relatively recent past that we have been able to cause significant harm to our environment, particularly the atmosphere. As a result, attempts to reduce air pollution and improve air quality date back less than 200 years.

Our ability to damage the atmosphere continues to grow more quickly than ever. The number and volume of atmospheric pollutants increases every year, and the effects of these pollutants have become increasingly destructive. Not only have we created chemicals that destroy the ozone layer and potentially disrupt our reproductive systems, but we now have the ability to change the actual make-up and functioning of the atmosphere.

Prior to the nineteenth century, air pollution was mainly caused by energy production for heating and cooking, such as wood or coal fires. In many parts of the world today, these continue to be significant sources of air pollution. In more developed countries, including Canada, motor vehicle emissions, coal fired electrical generation, and industrial and commercial sources have emerged as the main threats to air quality.

Attempts to curb air pollution through regulation date back to the Industrial Revolution in the mid- to late eighteenth century. During this era, governments were forced to take action against the damaging health effects of coal-burning industries often located in heavily populated urban areas. For the most part, early air pollution

As the sources of air pollution have become more numerous, complex, and interactive, tackling the problem is increasingly difficult.

laws were introduced by public health agencies at the local level. In Britain, *The Public Health Act* of 1848 imposed controls on smoke and ash emissions in the interests of public health. Similarly, in Canada, the 1884 Ontario *Public Health Act* provided for court action by local Boards of Health to abate odours, smoke, or dust if a hazard or potential hazard to health existed.[3]

Despite these laws, throughout the late nineteenth and early twentieth centuries, emissions from the burning of coal and other industrial processes became even more widespread. Transportation systems grew, and soon air emissions from trains (and later automobiles and aircraft) began to have a pronounced effect on the environment.

As time went on, new air quality issues emerged. The variety and quantity of chemical contaminants being used and released into the environment increased dramatically following the Second World War. Global production of organic chemicals doubled every seven or eight years since the 1930s, from approximately 1 million tonnes per year in that decade to 7 million tonnes in 1950, 63 million in 1970, and approximately 250 million in 1985.[4]

As the sources of air pollution have become more numerous, complex, and interactive, tackling the problem is increasingly difficult. In recent decades, we have begun to realize that local, regional, and even national efforts to address air pollution are not enough. Recognition of the transboundary and even global effects of some pollutants has led to international agreements to reduce key pollutants, such as ozone depleting substances and precursors to both acid rain and smog (see Chapter 2). These international agreements are critical to the long-term success of local, regional, and national air quality management efforts (see Chapter 3).

The Canadian Experience

Until the 1960s, air pollution in Canada was regulated primarily at the local level by municipal governments and public health boards. However, as air pollution sources became larger and more pervasive, it became clear that action was required at all levels of government.

Provincial efforts to control air pollution began in 1963, when Ontario became the first province to establish a permitting scheme for new industrial air pollution sources.[5] Several years later, Ontario passed the *Air Pollution Control Act* (1967), Canada's first comprehensive air pollution law. Over the next decade, a number of other provinces followed Ontario's lead.

Federal efforts to control air pollution began in the early 1970s. Several statutes were passed by Parliament, giving the federal government powers to control major sources of air pollution. For example, the government passed the *Motor Vehicle Safety Act* in 1970, providing it with authority to control motor vehicle emissions across the country.

The 1971 *Clean Air Act* was the federal government's first major attempt to manage air quality on a national basis. The Act authorized the newly formed Environment Canada to enforce regulations in two situations: where pollution sources presented a "significant danger" to the health of Canadians, and where a danger existed of violating an international treaty.[6]

In addition to these statutes, a number of other initiatives to reduce air pollution have been undertaken in Canada. Perhaps the best known is the 1985 *Eastern Canada Acid Rain Program.* Based on scientific evidence that acid rain was destroying lakes and trees across much of eastern Canada, the seven easternmost provinces agreed to reduce their emissions of sulphur dioxide (one of the main contributors to acid rain) by 50% by 1994. In return, the federal government contributed $150 million for pollution control technology and negotiated the 1991 *Canada–U.S. Air Quality Accord*, which led to significant reductions in transboundary flows of acid rain.

In recent years, there have been some important air quality

> Achieving political and legal changes necessary to significantly reduce air pollution will depend heavily on informed and effective public participation.

success stories. Trees and vegetation have begun to grow again around the smelters of Sudbury and other industrial towns; Vancouver's False Creek is no longer surrounded by smoke-belching woodwaste burners; and lead levels have declined significantly across the country.

However, new challenges continue to emerge. Acid rain still threatens 14,000 freshwater lakes in eastern Canada. Vancouver's urban air pollution problem has spread throughout the entire Fraser Valley. Certain airborne chemicals, known as endocrine disrupters, are believed to interfere with our reproductive systems. Clearly, more work remains to be done at all levels – in communities, provincially, nationally, and internationally.

Public Participation

Many Canadian initiatives to improve air quality, such as the 1985 *Eastern Canada Acid Rain Program*, have resulted from public concern and involvement. Achieving political and legal changes necessary to significantly reduce air pollution will depend heavily on informed and effective public participation.

However, a variety of institutional and systemic factors militate against such changes. These include lack of political will, scientific uncertainty, constraints on fiscal resources, and political/business opposition. Increasingly, therefore, citizens groups have begun to press for the right to play an enhanced role in environmental protection, through direct participation in policymaking and law enforcement.

One area where headway is being made concerns citizens' "right to know" about pollution levels and sources in their communities. The National Pollutant Release Inventory (NPRI) created in 1992 provides an important means for citizens to stay abreast of current emission information with respect to various air toxins.

Some progress has also been made towards securing legislative recognition of public participation rights. For example, in 1994 Ontario enacted an *Environmental Bill of Rights* that gives Ontario residents the right and ability to comment on some government decisions and policies affecting the environment, the right to request investigations where there is evidence that government has failed to enforce its environmental laws, and a limited right to sue to protect the environment "from significant harm." The 1999 amendments to the *Canadian Environmental Protection Act* include similar notice and comment provisions, as well as somewhat broader "citizen suit" provisions (see Chapter 6).

Even where such legislative rights are absent, citizens have increasingly been successful in persuading the courts to recognize their right to challenge government decisions that adversely affect air quality. A decade ago, Canadian courts were not accustomed to hearing, nor sympathetic to the goals of, such suits. Today, however, public interest environmental litigation is a well accepted and common feature of our court dockets, a reality that has been confirmed in a number of key 1990s legal precedents that affirm the role of the public in protecting the environment.[7]

Conclusion

Understanding and staying abreast of the science and law of air pollution is a significant challenge. This book is designed for those who are about to take on this challenge, as well as those who have been tackling it for some time.

Chapters 2 and 3 are aimed at relative newcomers to the air pollution issue. They are designed as primers on the scientific and legal dimensions of the air pollution problem.

Chapters 4 and 5 are designed to be accessible to newcomers, but also of interest and benefit to air quality activists. Chapter 4 discusses car emissions, in particular the various means by which we can reduce our dependence on motor vehicles and reduce the volume of harmful emissions. Chapter 5 then considers industrial source pollution, focussing on the traditional and emerging ways governments and other parties have sought to reduce emissions from so-called "stationary" sources.

Landmark Appeal

In 2000, the Sierra Legal Defence Fund argued a landmark appeal of permits that allow continued burning of wood waste material in beehive burners. Airborne particulate matter from these burners is an extremely serious health issue in northern BC communities. Children, the elderly, and asthmatics are particularly prone to ill-effects from the smoke produced by these burners. Not only is the case significant from a scientific perspective, but it also marks the first time that the Charter of Rights equality guarantees have been invoked to protect the rights of these particularly susceptible groups under our pollution laws.[8]

The final component of the book is designed to provide you with the means to take action on air quality issues . Chapter 6 is a "citizen's toolkit" of suggestions on how to stay informed about air quality issues, get involved in decision-making processes relating to air quality, and understand and evaluate the legal remedies you might have to promote better air quality. There is also an Appendix containing a wide variety of electronic sources for you to consult for further information.

Endnotes

1 Ministers of the Environment, 1996, *The State of Canada's Environment*, at G-1. See web-site: www.mbnet.mb.ca/ccme/pdfs/cat_eng.pddf

2 Estrin, D., and Swaigen, J. (eds), *Environment on Trial: A Guide to Ontario Environmental Law and Policy* (Toronto, Ont.: Emond Montgomery Publications, 1993), at 465.

3 Garrod, S., and Valiante, M., *The Regulation of Toxic and Oxidant Air Pollution in North America* (CCH Canadian Limited, 1986), at 87.

4 United Nations Environment Programme, "Hazardous Chemicals," *UNEP Environment Brief* No. 4, at 2. See web-site: www.unep.org/unep/library/library.htm

5 Garrod, S., *supra*, at 87.

6 Harrison, K., *Passing the Buck: Federalism and Canadian Environmental Policy* (Vancouver, BC: UBC Press, 1996), at 70-71.

7 See, for example, *Friends of Oldman River Society* v.*Canada*, [1992] 1 S.C.R. 3 and *R.* v. *Hydro Quebec*, [1997] 3 S.C.R. 213. The full texts of these decisions are available at the Supreme Court of Canada web-site at: www.scc-csc.gc.ca

8 "Sierra Legal Targets Air Polluters" (posted March 20, 2000 at www.sierralegal.org); see also "Beehive Burners: Putting Profits Ahead of Human Health," (SLDF, 1998).

a scientific primer

chapter 2

The World Health Organization recently stated that there is no safe level of human exposure to ground-level ozone. In other words, any amount of exposure can be harmful to humans.

Introduction

From the local to the most global, this chapter provides an overview of some of the key air issues of concern to Canadians today. Many of these issues are local, regional, or transnational in scope, including smog, acid rain, and hazardous air pollutants. Others are global, such as ozone layer depletion and climate change. The nature, impacts, and causes of each of these issues are discussed, as are recent emission trends.

Air pollutants are emitted from various sources. These sources are usually categorized as either *mobile* (cars, trucks, boats, and so on) or *stationary*. Stationary sources include emissions from specific *point* sources (industrial facilities, electrical generation plants, and incinerators) and emissions from geographically dispersed *area* sources (solvents, wood stoves, and home heating).

Smog

Smog refers to the yellowish fog that often envelops urban centres such as Toronto and Vancouver. It consists of a mix of pollutants, including ground-level ozone and particulate matter (primarily fine particulates, or particles less than 10 microns in diameter).

Ground-level ozone

Ground-level ozone is a pollutant that forms when two types of pollutants (volatile organic compounds and nitrogen oxides) react in the presence of sunlight, especially in warm weather conditions. In other words, ground-level ozone is not emitted directly into the atmosphere by any one source, but results from the *interaction* of two different types of pollutants. Because it is concentrated at or near ground level, it is sometimes referred to as tropospheric ozone. In Canada, ground-level ozone episodes occur primarily during the summer months.

Nitrogen oxides (NO_X) are produced by fossil fuel burning, primarily from motor vehicles, coal-fired electricity generation, and certain industrial processes (such as industrial boilers). Mobile sources account for 35% of total Canadian NO_X emissions in 1990. Electricity generation and industrial sources contribute 12% and 23% respectively, while other sources (including residential and commercial heating) contribute an additional 30%.[1] There are no significant natural sources of nitrogen oxides.

Volatile organic compounds (VOCs) are produced both naturally (biogenic emissions) and as a result of human activity (anthropogenic emissions).

Most human-made VOCs are produced as a result of fossil fuel use and distribution, including combustion in engines, fuel evaporation (e.g., at gas stations), and fumes from organic chemicals used in petroleum refining and other manufacturing processes.[2] VOCs also come from paints, solvents, and dry-cleaning fluid. In addition, VOCs are emitted into the air through the burning and decomposition of organic material such as wood and biological waste (e.g., sewage treatment and landfill waste decomposition).

Ground-level ozone causes serious health and environmental effects. The World Health Organization recently stated that there is no safe level of human exposure to ground-level ozone. In other words, any amount of exposure can be harmful to humans. Health effects related to ground-level ozone include increases in respiratory problems, medication use, physician and hospital emergency department visits, and hospitalization. Ground-level ozone can also lead to premature death. People with pre-existing respiratory or cardiac problems, children, and the elderly are most at risk.

Ground-level ozone has other harmful effects. For example, smog diffuses light, thereby reducing visibility and creating an unsightly yellowish haze over many urban centres. In addition, ozone can damage crops and natural forests.[3] Agricultural crop losses due to ground-level ozone exposure in the US are estimated at $2 to $3 billion per year. [4]

Health effects related to ground-level ozone include increases in respiratory problems, medication use, physician and hospital emergency department visits, and hospitalization.

Particulates

Particulate matter (PM) is a family of airborne particles from a range of sources, including smoke, dust, and pollen. These particles may be solid or liquid, organic or inorganic.

An important factor in determining the behaviour and impacts of particulate matter is the size of the particles. Particles range in size from 100 microns (µm) to 0.05µm (human hair is about 100µm in diameter). While early air quality monitoring and control efforts focused on total suspended particulate (TSP), scientists have increasingly begun to focus on the effects of the smallest members of this family.

Smaller members include what are known as coarse and fine particles. Coarse particles, known as PM10, are particles with a diameter of 10µm or less. Fine particles have a diameter of 2.5µm or less, and are known as PM2.5. Fine particulates can remain airborne for days or weeks and may be dispersed hundreds or even thousands of kilometres from their source.

While there are many natural sources of fine particulate (such as windblown dust or soil, pollen, and spores), those of most concern to human health are particles that are emitted directly into the atmosphere or that form as a result of chemical reactions in the atmosphere. These particles generally result from human activities such as fossil fuel combustion, industrial smelting, and waste incineration.

Particulate matter can have serious impacts on human health. Fine particulates can be inhaled into the depths of human lungs, which can aggravate asthma and other respiratory illnesses. The incidence of PM10 has been linked to both hospital admissions and mortality associated with acute respiratory diseases. As with ground-level ozone, children, the elderly, and those with cardiac or respiratory problems are particularly vulnerable. In addition, fine particulates can absorb toxic organic compounds and deliver them deep into the lungs where they are absorbed into the blood stream.

Particulate matter can also have serious environmental impacts. For example, it interferes with visibility and may often have strong and annoying odours. In addition, some types of particles are able to damage vegetation, including crops and forests. This is especially true of acidic aerosols, which are capable of increasing soil acidity, resulting in vegetation mortality.[5]

Acid Rain

Acid rain refers to any form of precipitation (including rain, snow, hail, or fog) that is highly acidic as a result of certain air pollutants. Acid rain is produced when sulphur dioxide (SO_2) and nitrogen oxides (NO_x) react chemically in the atmosphere to form sulphuric and nitric acid. These acids mix with water in the atmosphere and acidify soil and groundwater when precipitation occurs.

Once in the atmosphere, acid pollutants can be transported long distances on atmospheric pathways. For example, over half the acid rain in eastern Canada is from US sources.

Acid rain is caused by emissions of sulphur dioxide and nitrogen oxides. Sulphur dioxide, a colourless gas with a strong odour, is produced primarily by industrial processes such as metal ore smelting, petroleum refining, and cement manufacturing.[6] In addition, Canadian gasoline currently contains, on average, 343 parts per million (ppm) of sulphur, with Ontario's gas containing upwards of 540 ppm. This is the highest gasoline sulphur content in the industrial world (see Chapter 4).[7]

The transportation sector is the largest source of NO_x emissions in Canada. Other sources include coal-fired electricity generation, industrial boilers, and residential heating systems.

Acid rain affects soil and inland water. When soil and water are affected, entire ecosystems may be damaged. Acids also cause nutrients – such as calcium, magnesium, and potassium – to be leached from the soil. Loss of soil nutrients has been observed in Ontario and Quebec, resulting in reduced forest growth and higher vegetation mortality.[8] Acids also destroy beneficial soil micro-organisms.

Acid rain has also had a profound effect on freshwater lakes in

> Acid rain refers to any form of precipitation (including rain, snow, hail, or fog) that is highly acidic as a result of certain air pollutants.

Canada, particularly in the eastern part of the country. An estimated 142,000 lakes have suffered from acidification since the 1970s.[9] Despite significant progress in reducing SO_2 emissions, 11% of lakes in eastern Canada continue to acidify.[10] In the Atlantic region, acidification of spawning waters has caused a loss to the salmon fishery of 9,000 to 14,000 fish per year.[11]

Hazardous Air Pollutants

Hazardous air pollutants (HAPs), also referred to as air toxics, is a broad category that includes:

- polycyclic aromatic hydrocarbons
- "heavy" metals such as lead, arsenic, and mercury
- persistent organic pollutants (including PCBs and pesticides such as DDT, which is now banned in Canada).

HAPs are emitted by a wide variety of sources. These include fossil fuel combustion (e.g., in motor vehicles and industries), wood stoves, commercial chemical production and pesticide manufacturing, gasoline marketing, solvent use, and waste oil disposal.[12] Some toxins are also produced during waste incineration.

HAPs can cause immediate or long-term adverse effects on human health. For example, exposure to polycyclic aromatic hydrocarbons (PAHs) has been linked to cancer, growth retardation, and skin and eye disorders.[13] Another HAP, benzene, is thought to cause a form of leukaemia in humans.

Another form of HAPs are Persistent Organic Pollutants or "POPs." POPs are toxic substances that can take decades to break down or decompose (in other words, they *persist* in the environment). They also accumulate in the fatty tissues of fish, birds, carnivores, and humans. The toxic effects of these substances magnify as they move up the food chain, reaching very high concentrations in large mammals at the top, including bears and humans.

POPs can also be transported long distances over air and water. Air toxics, such as PCBs originating from as far away as Central America, often settle in cooler environments such as the Canadian Arctic. As a result, significant efforts have been made internationally to phase out the use of these pollutants.

New scientific research suggests that some POPs may mimic human hormones, interfering with reproduction and other developmental processes. Women and children are thought to be especially vulnerable to these substances, which may be transferred through breast milk or the placenta.[14]

Stratospheric Ozone Depletion

Most of the pollution discussed so far exists in the earth's atmosphere. Increasingly, however, human activity is causing damage to the stratosphere.

The ozone layer, which is found in the earth's stratosphere, contains relatively high concentrations of ozone molecules. This layer acts as a natural sunscreen, protecting the earth's surface from the sun's harmful ultraviolet (UV) radiation. However, the ozone layer has been getting thinner over the last few decades as a result of ozone-depleting substances emitted into the atmosphere.[15]

When ozone-depleting substances enter the stratosphere, UV light breaks them apart. The chemicals chlorine and bromine are released by this reaction. Molecules of these chemicals destroy ozone (made up of three oxygen atoms) by breaking one oxygen atom from each compound. After destroying one ozone molecule, the chlorine or bromine comes away unaffected. As a result, one single chlorine or bromine molecule is able to destroy thousands of ozone molecules. Furthermore, chlorine and bromine are not washed back to earth by rain like many other chemicals released into the atmosphere.[16] Instead, they are able to drift up to the stratosphere where they remain for twenty to 120 years or more.

Ozone-depleting substances are members of a large class of chlorine- and bromine-containing compounds, also known as industrial halocarbons. The primary ozone-depleting substances are chlorofluorocarbons (CFCs). CFCs account for more than 80% of

When the ozone layer is depleted, life on the earth's surface is exposed to increased UV radiation.

total stratospheric ozone depletion.[17] Other industrial halocarbons include halons, carbon tetrachloride, methyl chloroform, hydrochlorofluorocarbons (HCFCs), and methyl bromide. These substances are commonly used in refrigerators, fire extinguishers, and air conditioners. CFCs, the most common of these substances, are currently used as solvents to clean microchips and as degreasers and cleaners.[18]

When the ozone layer is depleted, life on the earth's surface is exposed to increased UV radiation. One type of UV radiation, UV-B, mutates genetic material in the cells of plants and animals. This leads to increases in skin cancer, suppression of the human immune response, reductions in crop yields, and death of phytoplankton – the foundation of marine ecosystems.[19] UV-B radiation can also cause eye damage and cataracts.[20]

Depletion of the ozone layer may also indirectly affect the global climate. Phytoplankton stores CO_2, an important greenhouse gas. The loss of phytoplankton, therefore, releases CO_2 from storage, thereby accelerating the rate at which CO_2 levels build in the atmosphere.

Global Climate Change

The atmosphere has a natural ability to insulate the earth's surface from heat loss. Naturally occurring greenhouse gases, such as water vapour, CO_2, and methane, trap some of the sun's heat and prevent its escape into space. This keeps the average global temperature of the earth at approximately 15°C. If there were no natural greenhouse effect, the average surface temperature would be about 34°C colder than it is today. However, human-related emissions of greenhouse gases, primarily CO_2, are changing the naturally occurring concentration of greenhouse gases in the atmosphere. Climate change refers to alterations in the global climate that are directly or indirectly caused by human activity.

Global climate has already started to change. Global average temperatures have risen 0.6°C in the last century, almost certainly as a result of human greenhouse gas emissions.

Greenhouse gases remain in the atmosphere for long periods of time, ranging from decades to thousands of years. As a result, it is difficult to reverse the effects of climate change. The Intergovernmental Panel on Climate Change , a body representing international scientific consensus on climate change, predicts that mean global temperature will increase by between 1.0°C and 3.5°C by 2100 if atmospheric concentrations of CO_2 are allowed to double. At present rates, this doubling could occur within the next sixty years.[21] This may not sound like much, but global average temperatures have only changed by 4°C since the last ice age, when Canada was under a kilometre of ice.

CO_2 is the major greenhouse gas. Human activity has caused a 30% increase in atmospheric concentrations of CO_2 since the Industrial Revolution, leading to the highest atmospheric concentrations of CO_2 in the last 240,000 years.[22] Other greenhouse gases include methane, nitrous oxide, CFCs, hydrofluorocarbons (HFCs), perfluorocarbons (PFCs), and sulphur hexafluoride (SF_6). HFCs are a common substitute for ozone-depleting CFCs but, like PFCs and SF_6, are a highly potent greenhouse gas that can persist in the atmosphere for hundreds if not thousands of years.

Canada is the second highest per capita greenhouse gas emitter in the industrialized world. This is largely the result of the energy-intensive lifestyles of many Canadians, which in turn is partly caused by our cold climate, the great distances between population centres, low-density suburban sprawl, and the production of fossil fuels for export.[23]

Canadians mainly rely on greenhouse-gas producing fossil fuels for transportation and home heating. Industrial and agricultural sources of greenhouse gases include energy production, power production, industrial processes such as smelting and cement manufacture, agricultural practices (including cattle production and the use of nitrogen fertilizers), and decomposition of organic matter in landfills. (For a discussion of some of the strategies to reduce greenhouse-gas emissions see chapter 4.)

While it is impossible to predict the precise rate and regional impacts of climate change in Canada, scientists have developed sev-

Canada is the second highest per capita greenhouse gas emitter in the industrialized world.

eral climate models. These models project temperature increases in most areas of Canada, particularly in the north. For areas prone to smog, increases in average temperature will mean an increase in the frequency and duration of smog episodes.[24] Other areas, especially Newfoundland and the Grand Banks, are projected to cool slightly. The hydrological cycle is also projected to change significantly, with increased winter rainfall in the west and summer drought in the Prairies.[25]

Extreme weather events are also expected to rise in Canada according to these models. Severe storms such as the 1998 ice storms in the Montreal area and eastern Ontario are examples of the types of events that are likely to increase in frequency.[26]

Climactic changes will also disturb delicately balanced ecosystems. For example, warming in Canada's north may result in loss of habitat for the polar bear. Forests, highly dependent on climatic conditions, may change dramatically, particularly in the boreal regions of Canada. Canada's coastal regions may lose land and be subjected to flooding as ocean levels rise. Populations of disease-carrying pests, such as mosquitoes, could thrive in the wetter, warmer climates.[27]

Emissions Trends

Progress towards reducing emissions of major air pollutants has been mixed. Some pollutants – such as SO_2, lead, and CFCs – have declined significantly over the last ten years, although SO_2 emissions have begun to rise again in Ontario (due largely to increased coal-fired electrical generation after the shutdown of several nuclear power stations in 1996). Others – particularly CO_2 and fine particulates – have climbed and are projected to keep rising over the next few decades. Emissions of NO_x, VOCs, and many toxic substances have remained relatively constant.

Success in reducing pollutants has often depended on the nature of the sources producing these emissions. For example, SO_2 reductions were achieved largely by controls on electricity generation utilities and metal smelting facilities in the seven easternmost provinces. Significant reductions were made possible by targeting a relatively small group of large emitters.

In contrast, transportation-related emissions of greenhouse gases have been extremely difficult to address. This is due to exponential growth in the number of vehicles on the roads, cars being driven further, and overall declines in the fuel efficiency of new vehicles. Similarly, emissions from energy production and use – the prime contributor to greenhouse gas and other key pollutants – have traditionally been difficult to reduce as they reflect overall patterns of high energy consumption by Canadians.

Endnotes

1 Environment Canada, 1996, "Acid rain," *National Environmental Indicator Series, SOE Bulletin* No. 96-2, Spring. See web-site: www1.ncr.ec.gc.ca/scripts/query2.exe

2 Mellon, M., Garrod, S., Ritts, L., and Valiante, M., *Regulation of Toxic and Oxidant Air Pollution in North America* (CCH Canadian Limited, 1986).

3 World Resources Institute, 1995, *The 1994 Information Please-Environmental Almanac*, at 116. See web-site: www.wri.org

4 US EPA Publication, "Ozone: Good Up High - Bad Nearby," July 1997. See web-site: www.epa.gov/ncepihom/catalog.htm

5 Environment Canada, Natural Resources Canada, Transport Canada, "Phase 2, federal smog management plan," 1997. See web-site: www.ec.gc.ca/phase2/execsum_e.htm

6 Environment Canada, "Acid rain," *supra.*

7 Mittelstaedt, M., "Auto, petroleum industries face off over cleaning Canada," *Globe and Mail*, 27 May 1998, at A3.

8 Environment Canada, "Acid rain," *supra.*

9 Environment Canada, *The State of Canada's Environment*, 1996, at 10-35. See web-site: www.ec.gc.ca/report_e.htm

10 Environment Canada, "Acid rain," *supra.*

11 CANDO (The Movement for Clean Air Now), "Outdoor air quality," 1997. See web-site: www.nb.lung.ca/body-312.htm

12 *Ibid.*

13 Hilborn, J., and Still, M., *Canadian Perspectives on Air Pollution* (Ottawa, Ont.: Environment Canada, 1990).

14 See, for example, Birnbaum, L.S., "Endocrine effects of prenatal exposure to PCBs, dioxins, and other xenobiotics: Implications for policy and future research," 1994, *Environmental Health Perspectives*, Vol. 102:676-679; or Colborn, Saal, and Soto, "Developmental effects of endocrine-disrupting chemicals in wildlife and humans," 1993, *Environmental Health Perspectives*, Vol. 101, No. 5.

15 Environment Canada, 1998, "Ozone depletion: A primer on ozone depletion," *The Green Lane*. See web-site: www.ec.gc.ca/science/splash.htm

16 *Ibid.*

17 *Ibid.*

18 *Ibid.*

19 *Ibid.*

20 *Ibid.*

21 Rolfe, C., *Turning Down the Heat: Emissions Trading and Canadian Implementation of the Kyoto Protocol* (Vancouver, BC: West Coast Environmental Law Research Foundation, 1998). See web-site: www.wcel.org/wcel/pub/1998/12248.html Also see the Intergovernmental Panel on Climate Change (IPCC) at www.ipcc.ch/ (copies of their publications are available at this site).

22 *Ibid.*

23 Environment Canada, "Climate change," *National Environmental Indicator Series, SOE Bulletin* No. 95-2, Winter 1995 update.

24 Last, J., Trouton, T. & Pengelly, D., *Taking Our Breath Away*, (David Suzuki Foundation: Vancouver, 1999)

25 Environment Canada, "Climate change," *supra.*

26 Environment Canada, 1998, "Global warming: A possible factor behind an increase in extreme weather events," *S&E Bulletin*, April/May 1998. See web-site: www.ec.gc.ca/science/splash.htm

27 *Ibid.*

a law and policy primer

chapter 3

The legal and policy framework governing air quality is almost as complex as the science of air pollution itself.

Introduction

Understanding the science and environmental implications of the air pollution problem is one thing; achieving the legal changes necessary to combat the problem is another. The legal and policy framework governing air quality is almost as complex as the science of air pollution itself.

The purpose of this chapter is to provide an accessible, general introduction to environmental law and policy, particularly as it relates to fighting air pollution. To this end, the chapter outlines the respective powers of different levels of government to take action against air pollution, provides an overview of how laws and policies are developed in Canada, and discusses some of the key factors that dominate public debate around air issues.

What Is Environmental Law and Policy?

Environmental law

Law has been defined as "the body of rules ... which a state or community recognises as binding on its members or subjects."[1] In Canada, with the exception of Quebec,[2] these laws are either part of the "common law" – law that has evolved through decisions made by judges over hundreds of years – or they are laws made by elected government representatives or legislators. This guide is primarily concerned with laws that are found in statutes (Acts of provincial and federal Parliaments) and regulations. It is these laws that establish the framework for air pollution control in Canada.

Statutes set out the powers of government agencies to protect the environment, as well as the rights and responsibilities of people relating to environmental quality. Most statutes are quite general in nature, and the precise details of when and how a law applies are dealt with by "regulation." Sometimes referred to as "subordinate legislation," regulations are as legally binding as statute law. Statutes normally give the power to make regulations to a designated Minister or Ministers of Cabinet.

For example, the *Canadian Environmental Protection Act* creates a framework for identifying and controlling toxic substances. A substance must be designated as toxic under the Act before the government can legally restrict its use. Both this designation decision and substance control measures are done by way of regulations made by the federal Cabinet.

In Canada, governments enjoy a broad discretion as to when and how they exercise their power to pass regulations. It is very difficult to legally challenge a government decision not to exercise its regulation-making power. Also, while statutes are subject to public scrutiny when debated in Parliament, regulations are often made with little or no prior notice to the public, let alone public input.

Environmental policy

While legislation provides the foundation for air quality management, the specifics of how (if at all) governments actually protect our air resources are often determined by policy.

Policies express a government's direction or goal with respect to a particular issue or problem. Governments develop policy for several reasons. Sometimes policy elaborates how a government intends to implement its legal duties and responsibilities. For example, the Canadian Environmental Assessment Agency has published a guide outlining how it interprets the *Canadian Environmental Assessment Act.* This guide is used by proponents of projects that are governed by the Act and by government departments to determine how they should meet their legal obligations under the Act (see Chapter 5).

Policies may also be used to elaborate government objectives or priorities that are not currently the subject of legislation. An illustration is the federal Pollution Prevention Strategy, originally

The question of jurisdiction – which level of government has authority to enact environmental laws or implement policies in a given area – is critical to understanding air quality management in Canada.

approved in 1994. When it first received approval, this strategy was aspirational, having no official status or force in law. Since that time, pollution prevention has been established as a legally mandated national goal, under the 1999 amendments to the *Canadian Environmental Protection Act*.

Another important illustration of environmental policy are the objectives set by governments with respect to *ambient air quality*. These objectives identify acceptable levels (or concentrations) of pollutants in the air. This information provides guidance to decision-makers as they issue permits, develop regulations and policies, or impose emergency plans to deal with unacceptable air quality. Under federal and provincial law, governments are empowered to establish air quality objectives. Ambient objectives are not, however, legally binding on politicians and bureaucrats. They merely represent targets or goals for pollution control efforts (see Chapter 5).

Jurisdiction: Who can make laws and policies?

The question of jurisdiction – which level of government has authority to enact environmental laws or implement policies in a given area – is critical to understanding air quality management in Canada.

It must be borne in mind that having jurisdiction does not mean a particular level of government is compelled to act to protect the environment; it merely establishes its ability should it choose to do so.

Law-making powers are granted to the federal and provincial governments under the Constitution. However, the *Constitution Act, 1867*, which delineates federal and provincial jurisdiction (or "heads of power"), does not allocate responsibility for the environment exclusively to either level of government. As a result, there are many areas of potential jurisdictional conflict in the sphere of environmental law and policy.[3]

When jurisdiction disputes arise, the courts are responsible for resolving them. In carrying out this function, they have increasingly sought to interpret the Constitution in a way that allows both levels of government to play a role in environmental protection.[4]

Provincial jurisdiction

Provinces get authority to manage air quality from two primary heads of power. The most important of these is the power to regulate on matters of *property and civil rights* within the province. It is under this power that provinces regulate air emissions and, more generally, waste management and environmental assessment.

Provinces also exercise constitutional power over *local works and undertakings*, as well as have ownership of most crown land. This power gives provinces the ability to regulate intra-provincial highways, railways, and other transportation links.

A key functional power of the provinces is the ability to create local government institutions and establish their mandates. For example, provinces can require local governments to regulate air emissions and adopt land use planning practices that minimize traffic and air pollution. In addition, provincial governments may delegate responsibility for air quality management to local government.

A case in point is the Greater Vancouver Regional District (GVRD), a regional partnership of twenty local government agencies in British Columbia. The BC government has delegated to the GVRD formal regulatory authority over industrial, commercial, and some residential sources of air pollution. The GVRD's air quality management program also includes extensive ambient air quality monitoring.

Federal jurisdiction

There are a variety of key federal powers relating to air quality management, including:

• The power to regulate *harmful conduct* by means of the criminal law, a power that has recently been construed by the

> To date, governments have been slow to employ their tax powers to promote a clean environment.

Supreme Court of Canada to support prohibitions to protect the environment.[5]

• The power to regulate matters of *national concern*, a power that has been used primarily to deal with transboundary and interprovincial environmental problems, including air quality.[6]

• Power to regulate its *own operations*. The federal government has used this power to require environmental assessment of projects undertaken by federal government departments or for which federal approval, federal funding, or transfer of federally owned land is needed.[7]

• Power to regulate emissions from *federal lands* (including military reserves, Indian reservations, and the two territories). The federal government has never used these powers to regulate air emissions but could do so.

• Power to regulate emissions from *federal works and undertakings* (interprovincial railways, trucking, and shipping companies). In addition, the federal government has the power to regulate *aeronautics and shipping*, including setting environmental standards for emissions.

Power to regulate air emissions in Canada's two territories has been partially delegated to the territorial governments. Such delegation can also occur under the terms of modern treaties entered into by the federal government with First Nations.

Notably, both the federal and provincial levels of government have jurisdiction to regulate air quality through their taxation powers. One way governments have used these powers is by exempting some "clean" motor vehicle fuels from generally applicable gasoline taxes (see Chapter 6). To date, however, governments have been slow to employ their tax powers to promote a clean environment.[8] For example, the federal government has resisted a broadly supported proposal to make employee transit passes, aimed at promoting use of public transit, tax-exempt employee benefits.[9]

Jurisdiction also has important implications for the field of international environmental law. The federal government is empowered to sign treaties on Canada's behalf. However, these treaties do not automatically become binding domestic law when signed. To become binding, the treaty must be incorporated into Canadian law in the form of either a federal or provincial statute or regulation. Historically, Canadian courts have said that the federal government has no special power to implement international law. Consequently, by ratifying a treaty, the federal government does not acquire the right to act or regulate on matters that would otherwise be within provincial jurisdiction.[10] More recent caselaw and commentaries, however, suggest that the federal treaty power may be broader and permit limited intrusions into areas of provincial jurisdiction.[11]

The main international air quality agreements to which Canada is a party include:

• **1979 United Nations Economic Commission for Europe Convention on Long-Range Transport of Air Pollution.** This agreement establishes a framework for regional cooperation on transboundary air pollution. A number of protocols have been negotiated under the Convention to reduce emissions of sulphur dioxide, nitrogen oxides, volatile organic compounds, heavy metals, and persistent organic pollutants.

• **1987 Montreal Protocol on Substances that Deplete the Ozone Layer.** This agreement commits signatories to reducing and eliminating the production and use of substances that deplete the ozone layer, most notably chlorofluorocarbons (CFCs). The schedule for implementing this agreement was subsequently "fast-tracked" by the 1989 *Helsinki Declaration on the Protection of the Ozone Layer.*

• **1991 Canada-United States Air Quality Agreement.** Initially focused on acid rain (including emission reduction targets for sulphur dioxide), governments have also begun to examine particulate matter and ground-level ozone under this agreement. Canada and the US are currently negotiating an ozone annex to the agreement.

• **1992 United Nations Framework Convention on Climate Change.** The purpose of this agreement is to stabilize greenhouse

Citizens are becoming more involved in environmental law and policy-making processes than ever before.

gas concentrations in the atmosphere at levels that do not cause dangerous interference with the climate system. Subsequently, *The Kyoto Protocol*, concluded in December 1997, commits developed country parties to the Convention on Climate Change to reduce their greenhouse gas emissions by 5% below 1990 levels between the years 2008 and 2012.

Intergovernmental bodies

As a result of overlapping jurisdiction, the federal, provincial, and territorial governments have begun increasingly to work through intergovernmental bodies such as the National Air Issues Coordinating Committee and the Canadian Council of Ministers of the Environment (CCME). These bodies do not have any official law-making authority. Despite this, they play an important policy development and decision-making role in relation to a wide variety of environmental issues, including acid rain, smog, and climate change.

This move towards negotiated intergovernmental decision-making has been criticized by many environmentalists. They argue that these processes have serious transparency and accountability deficiencies.

A focus of particular concern in this regard is the 1998 federal-provincial *Harmonization Accord*. The Accord includes a sub-agreement committing governments to develop "Canada-wide Standards" for a range of pollutants, including ground-level ozone and particulates (see Chapter 5). The sub-agreement states that where one level of government has agreed to act on a particular pollutant, the other order of government "shall not act in that role."[12] While neither the Accord nor the sub-agreement are legally binding, environmentalists contend that they will further deter the federal government from responding swiftly and decisively to emerging national environmental problems or challenges.

How Environmental Laws and Policies Are Made

Citizens are becoming more involved in environmental law and policy-making processes than ever before. This section describes how laws and policies are created and identifies opportunities for public input in each of these processes.

Legislation

Citizens can participate in the legislative process at various points and in various ways.

Before legislation is introduced into provincial Legislatures or the federal Parliament, government departments usually conduct extensive consultations, both internally within government and with interested non-governmental parties (often referred to as "stakeholders"). This process can often take many months. During this early stage government considers various approaches and goals and seeks input on the best way to achieve these goals.

When this process is completed, Cabinet (composed of elected members appointed by the Prime Minister or Premier) will be called on to approve the basic elements of the proposed new statute (or amendments to an existing statute). Once Cabinet approval is received, the proposal is drafted into the form of a bill, which is then put before the Legislature or Parliament. Before becoming law, a bill must undergo three readings. The first reading is the formal introduction of the bill. At second reading, the elected members debate and vote on whether to accept or reject the bill in principle. The federal Parliament and most provincial Legislatures then refer bills to a multi-party legislative committee, where they are examined in depth. Citizens can ask to make presentations to committees and can use this as an opportunity to suggest or support potential amendments.

The bill then returns to Parliament or the Legislature for third reading, where elected members debate the bill, propose and vote on amendments, and pass or defeat the bill. At the federal level, after third reading a bill must also go to the Senate for their consideration, a process that incorporates the same basic elements as in

The policy-making process offers an important opportunity for citizens wishing to get involved in air quality issues.

the House of Commons (i.e., three readings and consideration by a multi-party committee). Once the Senate has approved a bill, it is re-submitted to the House and signed by the federal Governor General. The Act then comes into force, usually as specified within the Act (often within ninety days).

Some statutes require an additional order of Cabinet before they come into force. Entry into force may be delayed or postponed for a number of reasons, such as the need to develop accompanying regulations or political resistance by interest groups. For example, the *Canadian Environmental Assessment Act* was passed by Parliament in 1992 but did not come into force until 1995. Similarly, the *Motor Vehicle Fuel Consumption Standards Act*, which would have resulted in stringent new vehicle fuel efficiency controls, was passed in 1981 but has never been proclaimed into force.

Policy

Policy both shapes and informs the interpretation of environmental law. Moreover, in areas where no legislative framework exists, policy serves as the primary basis for guiding government action. As such, the policy-making process offers an important opportunity for citizens wishing to get involved in air quality issues.

Policies may be developed by individual government departments (such as Environment Canada) or they may be government-wide initiatives. Departmental policies usually relate to issues for which the sponsoring department has sole responsibility or authority. Government-wide policies are used either for broad political priorities or for issues relating to the authorities and activities of more than one department.

Government-wide policies are ultimately made by the Cabinet. Where a government commands a majority in the House of Commons or Legislature and is united behind a policy initiative, chances of an initiative being blocked are very slim. At the same time, the policy-making process has become broader and more inclusive: it now often incorporates a wide range of groups and individuals including public servants, legislators, interested members of the public, and interest groups. As such, more opportunities exist to influence the direction and contents of particular policies.

Multistakeholder processes

Governments are increasingly using multi-stakeholder approaches to develop environmental law and policy. This approach brings together a number of interested parties, including government, industry, labour, and environmentalists. In theory, all are given an equal voice at the bargaining table.

In a multi-stakeholder exercise, government will establish a mandate for the process and then invite various stakeholders to participate. In most cases, multi-stakeholder groups are given the task of providing recommendations to government. Sometimes, government departments mandate an external group to develop solutions that the government then commits to implement.

The Canadian experience with multi-stakeholder processes has been mixed. These processes can take up a large amount of time, time that individual citizens may not be able to devote on an ongoing voluntary basis. They are also sometimes used by government to delay difficult decisions. In addition, working to achieve consensus may result in one group having an effective veto over all others. And lack of government will to act in the absence of consensus may create major barriers to effective processes.

Key Factors: Scientific Uncertainty and Economic Costs

Decision-making often involves finding the right balance between competing interests and views. Understanding some of the different factors at play in these decision-making processes can help citizens better prepare for their participation. Two key issues that often become the focus of debate with respect to air quality management are scientific uncertainty and economic costs.

A common feature in environmental policy debates, especially in those over air pollution regulation, is the issue of scientific certainty.

Scientific uncertainty

A common feature in environmental policy debates, especially in those over air pollution regulation, is the issue of scientific certainty. How certain must we be of the dangers posed by a pollutant to human health and the environment before we take action to control it?

The history of debate over leaded gasoline offers a vivid illustration of this point. When leaded gasoline was first introduced in the 1920s, it was vigorously opposed by many in the health community, who pointed to evidence about the adverse effects of lead dating back several decades. Despite these warnings, it was not until the 1970s, once a scientific consensus about its harmful effects had emerged, that lead was phased out and then banned by the Canadian government. Had a more precautionary approach been taken fifty years earlier, thousands of North Americans might have been spared serious neurological damages.[13]

Today, governments face similar questions about whether to act in the absence of absolute scientific proof. Examples of scientific uncertainty include the climate change debate[14] and the effects of the manganese gasoline additive MMT on human health and the environment.[15]

To avoid repeating the mistakes of the past, governments around the world adopted the *precautionary principle* in 1992. This principle, enshrined in the 1992 *Rio Declaration*, states that where there are threats of serious or irreversible damage, lack of scientific certainty is not a reason for governments to delay taking cost-effective measures to avert the potential harm. While the *Rio Declaration* did not bind subscribing governments, increasingly the precautionary principle has been incorporated in various international agreements and conventions. As a result, some have argued that it now forms part of customary international law.[16]

Economic Costs

Government policy-makers must also balance the *costs* against the *benefits* of potential actions to protect the environment. Historically, industry has often complained that the high cost of complying with environmental regulation will result in plant shutdowns and job loss. Confronted with these arguments, governments have sometimes retreated from implementing new environmental regulations or have relaxed enforcement of existing regulations.

Governments have begun to manage this issue in a number of ways. First of all, many government departments are required to produce cost-benefit studies prior to implementing new regulations or programs. These studies attempt to determine the potential costs of different options to help decision-makers make choices about appropriate new standards and approaches.

In addition, and perhaps more significantly, some governments and economists have begun to assess the economic *benefits* of taking action to protect the environment. This may include savings to the health care system as a result of improved air quality. For example, a Canadian government task group on acid rain calculated health and associated benefits of a 25% reduction in sulphur dioxide emissions in Canada and the United States. The group determined that savings in avoided health care costs were in the order of $210 million in Canada alone.[17]

An increasing number of economists also believe that environmental regulation can actually *stimulate* economic growth while protecting the environment. For example, Harvard University Professor Michael Porter has argued that carefully designed regulation can provide incentives for companies to innovate, allowing them to break from past practices and find new, cleaner, and more efficient production techniques.[18]

For regulation to play this role, Porter and others argue that society should regard pollution as a form of economic waste – a sign that materials are being used inefficiently and that resources are being wasted in handling, storing, and disposing of that waste.[19]

This implies moving from "end of pipe" regulation to an approach that is concerned with pollution prevention and "preventative design" (see Chapter 5). Through creative, incentive-based regulation, governments can encourage companies to implement innovative, environmentally sound, and cost-effective measures in the production process.

Endnotes

1 *The Concise Oxford Dictionary* (Oxford, Eng.: Oxford University Press, 1976).

2 Quebec, in contrast, is governed by the Civil Code.

3 Hogg, P., *Constitutional Law of Canada*, 2nd Edition (Toronto, Ont.: Carswell, 1985), at 598. See also *R.* v. *Hydro Quebec*, [1997] 3 S.C.R. 213. The majority decision holds that jurisdiction over "the environment" does not belong to either the provinces or federal government but is rather a shared responsibility.

4 *Friends of Oldman River Society* v. *Canada*, [1992] 1 S.C.R. 3

5 In the 1997 *R* v. *Hydro Quebec* case, the Supreme Court of Canada upheld federal regulation of toxic substances under the *Canadian Environmental Protection Act* on the basis of the federal government's power over criminal law. This decision appears to give the federal government a very broad ability to pass regulations for air quality protection if they form part of the regime enforcable under the criminal law.

6 *Interprovincial Cooperatives* v. *The Queen*, [1976] 1 S.C.R. 477.

7 *Friends of Oldman River Society, supra.*

8 For a novel and stimulating discussion of how governments can engage in "tax shifting" to promote environmental sustainability, see Durning, A., and Bauman, Y., *Tax Shift: How to Help the Economy, and Get the Tax Man off our Backs* (Seattle, Wash.: Northwest Environmental Watch, 1998).

9 See, for example, an initiative called "The National Task Force to Promote Employer Provided Tax-Exempt Transit Passes" on the following web-site: www.cutaactu.on.ca/advocacy.htm

10 *A.-G. Can.* v. *A.-G. Ont. (Labour Conventions)*, [1937] A.C. 326.

11 See *MacDonald* v. *Vapor Canada*, [1977] 2 S.C.R. 134 as discussed by Vanderzwaag, D., and Duncan, L., "Canada and Environmental Protection" in Boardman, R. (ed.) *Canadian Environmental Policy: Ecosystems, Politics and Processes* (Don Mills, Ont.: Oxford University Press, 1992), at 6.

12 Canadian Council of Ministers of Environment, 1998, *Canada-wide Accord on Environmental Harmonization - Canada-wide Environmental Standards Sub-Agreement*, Section 4.4. See web-site: www.mbnet.mb.ca/ccme/3e_priorities/3ea_harmonization/3ea2_cws/3ea2a.html

13 See *Rachel's Environment and Health Weekly*, Nos. 539 to 541 (Annapolis, Ind.: Environmental Research Foundation, 1997).

14 For an overview of the climate change debate, see Gelbspan, R., "The heat is on: the warming of the world's climate sparks a blaze of denial," *Harper's*, 31 December 1995. See also Calvin, W.H., 1998 "The great climate flip-flop," *The Atlantic Monthly*, January, 1998. For a more comprehensive overview of climate change, and what can be done to counter it, see Rolfe, C., *Turning Down the Heat: Emissions Trading and Canadian Implementation of the Kyoto Protocol* (Vancouver, BC: West Coast Environmental Law Research Foundation, 1988), or web-site: www.wcel.org/wcelpub/1998/12248.html

15 For a discussion of the debate surrounding MMT's effects, see McKinsey, V., "Running on MMT? The debate on the health effects of a gasoline additive rages on," *Scientific American*, June 1998. See web-site: www.sciam.com/1998/0698issue/0698techbus2.html

16 See, for example, Sands, P., *Principles of International Environmental Law*, 1995, at 208-212.

17 Environment Canada, 1998, "Health and atmosphere both benefit from reducing sulphur in gasoline," *S&E Bulletin*, February 1998. See web-site: www.ec.gc.ca/scence/splash.htm

18 Porter, M.E., and van der Linde, C. "Green and competitive: Ending the stalemate" (Sept - Oct 1995), *Harvard Business Review*, at 120.

19 *Ibid.*, at 122.

motor vehicle air pollution

chapter 4

Motor vehicle air pollution has become a seemingly inescapable reality of life in urban Canada.

Introduction

Motor vehicle air pollution has become a seemingly inescapable reality of life in urban Canada. Pollutants emitted from motor vehicles include volatile organic compounds (VOCs), nitrogen and sulphur oxides (NO_X and SO_X), greenhouse gases, carbon monoxide (CO), hazardous air pollutants (HAPs), particulate matter (PM), and ozone depleting substances. In Canada, direct tailpipe emissions from cars and light trucks account for the largest proportion of air pollutants. But other forms of transport – air, truck, marine, and rail – are also significant contributors to air pollution. In addition, oil and gas production and refining also have serious negative impacts on air quality.

While motor vehicles emit far less pollution per kilometre today than twenty years ago, a number of factors have meant that the quality of our air has continued steadily to deteriorate: our population has grown, per capita car ownership has increased, transit usage has gone down, and the numbers of trips per car and length of trips have grown. From 1950 to 1992, per capita car ownership increased by almost 300%. Kilometres travelled by urban automobiles increased five-fold while urban transit rides per person per year declined from 250 to under 100.[1] Moreover, greenhouse gas emissions per kilometre travelled by new passenger vehicles has increased by 13% over the last decade.

Technological fixes – cleaner cars, cleaner fuels, etc. – have the potential to yield significant reductions in motor vehicle pollution. But the long-term goal must be to reduce reliance on the vehicle because there is a limit to the environmental benefits technology can offer. For example, while it may be possible to develop cars that emit one-tenth of the greenhouse gases of today's vehicles, a car that is twenty or thirty times as efficient – the type of improvement that may be needed to achieve long-term climate protection – may not be possible. As well, many technological fixes solve one environmental problem but create another. Hydrogen fuel cells eliminate tailpipe emissions but, depending on how the hydrogen is produced, may be no cleaner than gasoline engines.[2] Similarly, the batteries used in some electric vehicle designs may lead to increases in our use of heavy metals.

Our cities and communities have been designed to accommodate roads and automobile use. Twenty-five percent of our urban space is devoted to pavement: parking lots, streets, and highways. While the car has given us a great deal of mobility, dependence on the car has perversely restricted our mobility. Congestion caused by too many cars leads not only to pollution but also to frustration. More highway space tends to provide only short-term relief of congestion; in the longer term more highways encourage urban sprawl and greater reliance on the vehicle. We need to fundamentally shift our current land use patterns to create communities and developments that are transit and pedestrian friendly, so that driving is no longer a necessity.

This chapter focuses on two themes:

- how we can reduce our dependence on the motor vehicle
- how we can reduce emissions per motor vehicle kilometre travelled.

Reducing Dependence on the Motor Vehicle

Reducing dependence on the motor vehicle is an essential strategy for reducing air pollution from personal transportation. This strategy requires a rethinking of the way our communities are designed and yields a number of tangible benefits: fewer traffic accidents, less land devoted to pavement, less land devalued by adjacent traffic, reduced noise pollution and vibration, and less water pollution from road run-off. The solutions also help reduce municipal taxes. Finally, they yield some less tangible but very real benefits: more

Reducing dependence on the motor vehicle is an essential strategy for reducing air pollution from personal transportation.

complete and people-friendly cities and a greater sense of community as streets become safer and calmer places for pedestrians, cyclists, and children.

The land use-transportation connection

Land use decisions profoundly affect transportation choices and therefore our air quality. All too often, as growth occurs, new developments involve construction of low-density suburban housing away from business and commercial districts. Business, commercial, and residential areas are separated from each other, forcing people to drive between them. Children are too far away from school or playgrounds to walk or cycle and end up becoming dependent on cars.

Such sprawl is expensive in comparison to compact developments. It costs more for municipalities to provide water, sewer, and road services to large residential lots, whereas more compact housing uses less land and can be designed to have dramatically reduced heating and energy costs.

Transit service is also affected in low-density suburban areas, as more buses are required to service fewer people. The high costs of running these transit systems result in lower frequency and quality of service, which in turn result in reduced customer use, as people prefer the convenience of their own cars. By contrast, higher density communities ensure that there are enough residents in a given area to sustain good transit routes. Transit becomes profitable and service levels increase to a point where the convenience of transit comes close to or even exceeds that of the single occupancy car.[3]

Municipal zoning laws also have an influence on the type of community that develops. Mixed- or multi-use zoning, rather than single-use zoning, allows for shops, businesses, and residences to be integrated. A well-planned mixed-use community is more compact, allowing people to walk, cycle, or use transit more easily. In higher density, mixed-use communities, air pollution can be quickly reduced because the increased walking, cycling, and transit trips replace the short, cold-engine car trips that pollute the most per kilometre.[4]

Similarly, choices concerning how transportation dollars are invested directly affect land use and air quality. Transportation engineers often argue that air quality is served by relieving congestion, which allows cars to travel more freely and efficiently. But this ignores the known effect that more roads lead to more kilometres travelled. Investments in transit, on the other hand, will improve service and therefore encourage use. Transit-oriented developments create communities around transit access and more compact development along transit corridors, therefore encouraging greater ease and convenience of transit, while potentially discouraging automobile traffic.

Creating complete communities and reduced car dependency

There are a number of steps municipal, regional, provincial, and federal governments can take to improve transportation choices, reduce reliance on the car, and encourage more complete urban developments. These measures work best when taken together.

Improved planning

Transportation infrastructure development and land use must be guided by a plan that considers long-term goals and benefits for the region and long-term costs of different transportation and land use choices. Alternative transportation programs and projects should be allowed to compete equally with roadway improvements on the basis of total costs. Steps can be taken on various fronts to ensure this happens, including:

- **Integrated least-cost transportation planning as a condition of federal-provincial transportation grants.** Intelligent transportation is essential to urban environmental quality and economic vitality. Least-cost planning means that strategies to reduce automobile use are considered equally with strategies to increase capac-

Intelligent transportation is essential to urban environmental quality and economic vitality.

ity. Instead of favouring road building and expansion, major new transportation projects have to be compared against alternatives. Environmental impacts of both proposed projects and their alternatives need to be considered. Bicycle and pedestrian facilities should be considered and included in all new transportation infrastructure. Provincial and federal transportation funding should be conditional on projects being part of integrated transportation plans that evaluate the total fiscal, environmental, and social implications of different options

- **Improved growth management legislation.** Provincial governments can adopt laws that require municipalities and regions to develop integrated planning for transportation and urban growth. These laws need mechanisms to ensure that all municipalities bring their zoning into compliance with regional plans and that regional plans are not reduced to the lowest common denominator.

- **Municipally led planning initiatives.** Even in the absence of provincially or federally mandated integrated least-cost transportation and land use plans, municipalities and regional districts can develop them. The plans should incorporate all of the components discussed next.

Changing land use regulation

Over the long term, municipalities and regional governments can dramatically enhance air quality by passing zoning bylaws that encourage compact growth patterns that are amenable to transit. The following measures will help ensure better transit-oriented land use:

- **Development standards.** Typically, zoning bylaws and requirements for new roads stifle innovative developments that are affordable, increase density, preserve greenspace, and support transit. Municipalities can adopt standards that mandate such developments and prohibit those that are unfeasible to service by transit. Municipalities can also adopt alternative development standards that give developers a choice between status quo patterns and sustainable alternatives.

• **Allowing mixed uses.** Zoning bylaws that allow for home businesses and a mix of residential and non-polluting business and commercial uses.

• **Allowing greater density.** A number of zoning restrictions limit the in-filling of existing suburban areas. Mandatory setbacks (i.e., minimum front, back, and side yards) and density restrictions can be relaxed to allow multiple-family dwellings and secondary suites while preserving or improving the ambience of neighbourhoods.

• **Reducing taxes on buildings and increasing taxes on land.** Currently, municipal taxes are paid on the assessed value of land and buildings. Decreasing tax rates payable on buildings and increasing rates payable on urban land would encourage more compact development as well as provide more affordable housing in the urban cores.

Parking management

Municipalities control the supply of parking spaces. Most zoning bylaws mandate minimum free parking requirements for new homes and businesses. Cities control the price and amount of on-street parking and set tax rates on off-street parking lots. The provision of free parking at businesses, shopping centres, and housing developments encourages single occupancy vehicle use. Moreover, free parking also has a financial price: businesses that are required to spend money on developing "free" parking incur costs that are reflected in the prices of goods and services.

There are a number of steps municipalities, employers, and senior levels of government can take to expose the true costs of parking spaces and reduce car use:

• **Eliminate parking requirements in zoning bylaws.** The rationale for mandatory provision of free off-street parking is to avoid shoppers and employees parking in residential areas, but passing

Acid rain has also had a profound effect on freshwater lakes in Canada, particularly in the East.

and enforcing residents' only on-street parking would be cheaper than providing free parking. [5]

- **Provide developers with alternatives to free parking.** Where municipalities are unwilling to deregulate parking, they could give developers increased options. For instance, developers could be given the choice of providing improved bicycle infrastructure and carpooling as alternatives to off-street parking requirements.

- **Eliminating the tax subsidy for free parking.** Currently, the federal *Income Tax Act* treats employers' provision of free employee parking as a non-taxable employee benefit. On the other hand, employees and employers pay tax on free transit passes. Ideally, the federal government should tax free parking benefits and provincial and federal employment legislation should give employees the right to receive pay in lieu of free parking. Alternatively, free transit passes should be non-taxable. [6]

- **Charge more for parking.** Municipalities can increase charges for on-street parking, increase the number of hours that parking charges are in effect, and increase the tax rates payable by privately operated parking lots. Dedicating revenue from parking meters to local improvements such as street furniture, bicycle facilities, and transit shelters will tend to increase the acceptability of higher rates.

Eliminating vehicle subsidies and increasing variable costs

Almost half the price of gas in Canada is tax. However, the tax revenue generated is exceeded by the costs of driving paid by taxpayers and not borne by the driver: the cost of land for roads, road building, road maintenance, air pollution, sprawl, traffic enforcement, and hospitalization of accident victims.[7] Not only are cars subsidized, but most of the costs of owning and driving one are fixed: insurance, license, registration, depreciation, and purchase or financing. Only a fraction of the costs are variable: gas, oil, tires, and maintenance. Increasing the variable costs of driving can influence how much people rely on their cars as opposed to public transportation. The following are some examples of the way subsidies can be reduced and variable costs can be increased:

• **Road tolls and fuel taxes.** Road tolls and fossil fuel taxes are useful means of eliminating vehicle subsidies. Tolls should be applied systemically – not to specific roads or bridges – so that they can encourage reduced travel at peak hours rather than the use of longer routes to avoid tolls.[8]

• **Pay-per-kilometre car insurance.** Currently, drivers pay an average of seven to ten cents per kilometre for auto insurance. Pay-per-kilometre insurance can be more actuarially accurate (i.e., increasing the proportion of insurance paid by drivers most likely to have accidents as they are driving more kilometres) and would provide an incentive to reduced vehicle use.[9]

Improving transportation infrastructure

Investments in transportation infrastructure affect people's transportation choices. Least-cost transportation planning will often lead to investments in the following:

• **Transit.** Often the most cost-effective investments are in simple buses and streetcars or in programs to make them more efficient. Vancouver's B-Line Bus is an articulated bus featuring limited stops and bike racks. Its introduction led to a 20% increase in ridership from car drivers along the bus corridor. In low-density areas, alternatives such as subscription van and carpools can be cost effective.

• **Trip reduction programs.** Programs that discourage single occupancy vehicle use and encourage alternatives can be implemented at a workplace or regional level. These include "carrots" such as increased transit, improved bicycle infrastructure, and reduced parking rates for carpools and vanpools, as well as "sticks" such as higher parking fees for single drivers.

Almost half the price of gas in Canada is tax. However, the tax revenue generated is exceeded by the costs of driving paid by taxpayers and not borne by the driver.

• **Dedication of existing road space to transit and high occupancy vehicles (HOVs).** Dedicating existing lanes to buses and HOVs can encourage carpooling and transit use.[10]

• **Transit priority measures.** A number of measures can be taken to increase the efficiency of transit and its competitiveness to the car. Bus bulges (curb extensions from the sidewalk) allow buses to stop, load, and unload without having to exit and re-enter the main flow of traffic. Buses and traffic lights can be linked by computer and radio to minimize unnecessary stops.

• **Building code requirements.** Provinces and municipalities can require all new office buildings to provide bicycle facilities such as secure bike lockers and showers.

• **Free transit.** Public transit can be offered free of charge in downtown cores and on days when urban pollution is very high.

• **Bicycle infrastructure.** Bike paths, bike lanes, marked bike routes, and cyclist-activated lights throughout the city will decrease urban traffic. In Vancouver, a newly developed bikeway saw a 300% increase in usage within its first year, and there is now 100 km of bikeways in the city.

• **Traffic calming.** Broader sidewalks, more trees along streets, slower speed limits on local streets, textured road surfaces, traffic circles, and diverters can all slow the speed of traffic on local and neighbourhood streets and ensure that cars keep to major thoroughfares. This can make cycling and walking more attractive, as well as adding to community safety and providing a higher quality environment for businesses.

Reducing Emissions per Kilometre Travelled

While reducing the number of kilometres driven is critical to combatting air pollution, equally pressing is the need to reduce the level of harmful vehicle emissions. Both the federal and provincial governments can take significant steps in this area.

Reducing emissions requires more than simply installing a catalytic converter or other control device. What is needed is a co-ordinated strategy encompassing the following elements:

- stricter vehicle emission and fuel standards
- enhanced fuel efficiency requirements
- tax exemptions for alternative and alternative fueled vehicles
- improved vehicle inspection and maintenance programs
- incentives to accelerate the retirement of higher polluting vehicles.

Vehicle emission standards

In Canada, both federal and provincial governments can increase the stringency of emission standards. Because Canada is integrated into the North American automobile market, any new regulations are likely to be closely tied to those in force in the US. This does not mean that Canada cannot adopt more stringent regulations, but ours are likely to be based on elements of US federal or state regulations.

US Tier 1 and Tier 2 standards

Traditionally, in the US vehicle emission standards have been set for different types and sizes of vehicles. Each standard sets limits on levels of polluting substances released per kilometre travelled. Manufacturers must then go through testing and certification procedures to ensure that all the model/engine combinations they sell meet the required limits. Currently, US federal regulations require all vehicles to meet what is known as the Tier 1 standard. As of 2004, manufacturers will be required to meet newly announced

While reducing the number of kilometres driven is critical to combatting air pollution, equally pressing is the need to reduce the level of harmful vehicle emissions.

Tier 2 standards, which are considerably more stringent.

California standards

California has adopted a different and more strict approach than the US federal government. Many US states are beginning to follow its lead. California allows manufacturers to sell a mix of vehicles certified to different emission classes. The mix is regulated by a vehicle "fleet average." The lower the fleet average, the cleaner the mix of vehicles a manufacturer is required to sell.

Canadian standards

Canada has tended to lag significantly behind the US in adoption of emission standards. Canada adopted Tier 1 standards in 1997. This had minimal impact on emissions as all but a few manufacturers were already implementing the US standards in Canada.

Whether Canada will follow the US lead and adopt Tier 2 standards as of 2004 is uncertain, although the Minister of Environment has announced that this is his goal.[11] BC is the only province to have up-to-date emission standards.[12] Currently, BC's standards are based on the California fleet approach. BC's regulation also includes a voluntary target for sales of clean technology vehicles. And BC has just announced it will adopt US Tier 2 emission standards as of 2004.[13]

Fuel standards

The constituents of gasoline, diesel, and other fuels have a major impact on human health and the environment when combusted. Depending on how fuel is formulated, emissions of SO_2, CO_2, NO_x, VOCs, PM, and HAPs vary.

The US government, California, and BC all have adopted comprehensive standards for reformulated gasoline (RFG). The Canadian federal government limits lead and benzene concentrations, and is proposing to limit sulphur as well. Currently, Quebec and Ontario only regulate vapour pressure. A table that summarizes gasoline standards in Canada is found at the end of this chapter.

Tightening gasoline standards can improve air quality dramatically. The US RFG standard in place as of 2000 will reduce tailpipe emissions of VOCs by 27%, HAPs by 21.5%, and NO_x by 6.8% as compared to 1995 standards. The cost of adopting stringent RFG standards is modest compared to the benefits. California estimates that its RFG requirement adds an average of 2.5 US cents per litre to gasoline.[14]

Sulphur in gasoline

One of the most contentious issues in relation to fuel quality has been sulphur levels in gasoline. At an average of 343 ppm, Canada's gasoline sulphur levels are among the highest in the industrial world.[15] Sulphur in gasoline significantly affects emissions of fine particulates, nitrogen and sulphur oxides, VOCs, and carbon monoxide and as such poses serious health risks. It has been estimated that if Canada adopted a stringent sulphur standard, between 2000 and 2020 we could expect to avoid 2,100 premature deaths, 6,800 emergency room visits, 3.3 million asthma symptom days, and 11 million acute respiratory symptoms.[16]

In October 1998, Environment Canada released a proposal for a 30 ppm standard starting in 2005, with an interim requirement of 150 ppm starting in 2002. This proposal has been criticized as too slow by the auto industry, health experts, and environmentalists, and too aggressive by gasoline producers. At time of print, the US EPA was expected to announce its intention to phase in a 30 ppm average requirement from 2004 to 2006. Gasoline refiners will likely argue that Canada should "harmonize" its standards with the slower US phase-in.

Dirty Gasoline

In 1998, the Sierra Legal Defence Fund went to court to force six major oil companies to release information on sulphur levels in their gasoline, which they were required to provide to Environment Canada. After fighting against disclosure for over a year, the companies finally gave in. The information reveals that gasoline refined in Ontario, particularly by Imperial Oil, is the dirtiest in Canada, with the worst culprit being the Esso refinery in Sarnia that produced regular grade gasoline with 810 ppm of sulphur in the summer smog season (July-September) of 1998.[17]

Sulphur in diesel

The federal government also regulates diesel fuel sulphur concentrations.[18] Since 1998, sulphur for on-road diesel fuel has been limited to a maximum of 500 ppm.

There is currently no regulation of sulphur in diesel fuel for off-road uses (off-road includes trains, boats, and machinery), which accounts for approximately 45% of diesel usage. Development of a sulphur standard for off-road diesel and more stringent on-road standards would yield major health benefits.[19]

Renewable fuel content requirements

Fossil fuels can be replaced with alcohol (ethanol or methanol) produced from renewable sources such as wood or grains. Potentially, this will reduce greenhouse gas emissions, as the carbon dioxide emitted by burning ethanol or methanol will be balanced by the carbon dioxide removed from the atmosphere during the growth of the wood or grain.

Since ethanol can be blended with gasoline and used in a regular engine, it is possible for government to mandate minimum renewable content requirements in gasoline. Canadian environmentalists have recommended phasing in a requirement that all transportation fuels have a minimum of 5% renewable content by 2010.[20]

Improving fuel efficiency

Canada can improve the fuel efficiency of new cars through both fuel efficiency standards and feebate systems.

Fuel efficiency standards

Fuel efficiency standards (also known as fuel consumption standards) set limits on the amount of fossil fuel a motor vehicle may burn per kilometre travelled. Fuel consumption is directly related to carbon dioxide emitted by vehicles, and adoption of more stringent standards is one of the single most effective measures government could take to reduce greenhouse gas emissions. Also, with less fuel being burned, polluting chemicals released by fossil fuel production and combustion decrease.

In the US, fuel efficiency standards for new cars and new light trucks are set under Corporate Average Fuel Efficiency Standards (CAFE). CAFE requires manufacturers to sell a mix of vehicles that, on average, meet fuel efficiency standards. In Canada, under legislation passed in 1981, the federal government was empowered to set fuel efficiency standards for motor vehicles; however, the law was never brought into force.[21] Instead, Natural Resources Canada set up voluntary agreements with the motor vehicle industry. Manufacturers agreed to provide vehicles that meet Company Average Fuel Consumption (CAFC) standards equivalent to the US CAFE standard.

CAFC standards have not improved since 1985. For cars, they are 8.6 litres per 100 km (27.5 miles per gallon). The voluntary standard for trucks is 11.8 litres per 100 km. Without the impetus of regulation, the fuel efficiency of new cars in Canada has not improved since 1990. Fuel efficiency of new light trucks (including minivans and sport utilities) has actually declined from 10 litres to 11.4 litres per 100 km. Because of this trend and the increased sales of sport utilities, minivans, and trucks, the overall efficiency of new vehicles (both cars and trucks) sold in Canada declined by 13% between 1986 and 1997.

Environmentalists and the British Columbia Automobile Association have both called for legally binding stringent fuel efficiency standards for Canada. The West Coast Environmental Law

> Fossil fuels can be replaced with alcohol (ethanol or methanol) produced from renewable sources such as wood or grains.

Association has recommended implementation of a 6 litre per 100 km standard for passenger cars and light trucks by 2005.

More stringent efficiency standards are technologically feasible. According to the US-based Rocky Mountain Institute, within a decade cars could be made ten times more efficient with no loss to handling or passenger capacity and no increase in price.[22] Hybrid vehicles like the Toyota Prius, introduced in Canada in 2000, reduce fuel consumption to 3.6 litres per 100 km. The Mitsubishi Gallant, which is not available in Canada due to the high sulphur content of our fuel, consumes only 3.15 litres per 100 km using advanced fuel injection technology.

Improved fuel efficiency is also economically worthwhile for reasons unrelated to climate change or air pollution. The Climate Action Network estimates that phase in of a stringent standard in Canada would save Canadians $4 billion in reduced gasoline bills over and above any increased vehicle costs. The Pembina Institute for Appropriate Development has estimated that any increased costs to cars would be paid back in fuel savings within three to four years.[23]

It should be emphasized that fuel efficiency standards must be coordinated with efforts to reduce vehicle miles travelled. With lower gas consumption and thus lower gas bills, there is evidence that consumers drive more, reducing the effectiveness of the efficiency standard. For instance, a 10% increase in efficiency may lead to a 1-3% increase in kilometres travelled.[24]

Feebates and gas guzzler taxes

Governments are increasingly experimenting with *feebates* (under which cars purchasers receive a subsidy or pay a fee depending on the fuel efficiency of the car they are buying) and *gas guzzler taxes* (which levy a surcharge on purchasers of inefficient vehicles). Both feebates and gas guzzler taxes appear to be effective, although their effectiveness has often been hampered by poor design and implementation. For instance, in the United States, a gas guzzler tax of between $1,000 and $7,700 is applied to gas guzzling passenger cars, but the tax is levied on less than 2% of sales and is hidden in the purchase price.[25] Despite this, it appears the tax has helped to improve the fuel economy in luxury cars.[26]

Ontario has had a feebate system in place for years. However, several factors have limited its effectiveness: 90% of cars are not subject to either the surcharge or the subsidy; the maximum surcharge or subsidy is a small faction of the purchase price; and most consumers only learn of the tax after they have decided to buy a car.[27]

Feebate systems could be made significantly more effective by increasing consumer awareness and making the fees variable across the whole range of fuel efficiencies. A 1993 analysis showed that a feebate system alone could stabilize passenger vehicle emissions, leading to a more fuel efficient vehicle stock more quickly and at less cost than a gasoline tax.[28] Although there may be increases in vehicle costs, consumers would save due to reduced energy costs.[29]

Tax exemptions for alternative fuels and alternative fueled vehicles

The federal and many provincial governments exempt some alternative fuels from fuel sales and excise taxes. British Columbia, for instance, exempts propane, natural gas, and fuels that contain over 85% ethanol or methanol. The provincial exemption is worth the equivalent of 11-15 cents per litre of gasoline and costs the provincial treasury $25 million per year.[30] The federal government exempts the same fuels from its fuel excise tax.

Proposals have also been made for tax deductions for alternative fueled vehicles. These exemptions have been justified on the basis of their lower emissions potential. Vehicles manufactured to run on natural gas and propane tend to have lower emissions of NO_X, VOCs, PMs, and HAPs.[31]

However, tax exemptions should be carefully considered. First, reduced fuel prices can encourage increased mileage, and thus greater overall emissions. Second, vehicles converted to run on

Improved fuel efficiency is also economically worthwhile for reasons unrelated to climate change or air pollution.

propane and natural gas tend to be big polluters. Forty-nine percent of propane fueled vehicles fail BC's AirCare emission inspection program and 29% of natural gas vehicles fail. These are the overwhelming majority of the vehicles that take advantage of the tax exemption.[32]

Keeping vehicles clean: Inspection and maintenance programs

Vehicle inspection and maintenance (I/M) programs require drivers to have their cars and trucks periodically tested for emissions and also require the repair of excessive emitters.

Both BC's AirCare program and Ontario's Drive Clean program require all light-duty vehicles within certain geographic areas to pass an annual emission inspection as a condition of licensing. AirCare applies to the lower Fraser Valley; the first phase of Drive Clean applies to Greater Toronto and Hamilton-Wentworth. In October 1999, Drive Clean expanded to cover heavy-duty vehicles and by 2002 it will apply to all of southern Ontario.

The effectiveness of I/M programs depends on a number of factors, including:

- **Stringency of standards.** Although vehicles are certified to meet new vehicle emission limits for 50,000 miles of use, I/M programs usually impose less stringent limits. Current AirCare tests allow emission levels five to ten times the original manufactured standard, and over 86% of vehicles pass. Tightening the limits could further reduce emissions.

- **Conditional passes.** Both AirCare and Drive Clean offer conditional passes for vehicles where repair costs exceed a certain amount. This allows car owners to spread the cost of repairs over time. Restricting owners to a one-time conditional pass would increase the effectiveness of I/M programs.

• **Warranty requirements.** Manufacturers currently warrant emission control equipment, but may not warrant that a vehicle will pass Canadian I/M tests. The number of conditional passes can be limited by requiring manufacturers to warrant that their vehicles will pass the I/M test and not offering conditional passes for vehicles under warranty.[33]

• **Centralized tests only.** Under AirCare, vehicles are tested at a number of centralized test facilities located around the Lower Mainland. Centralized testing separate from repair facilities ensures that testers have no interest in recommending unnecessary or expensive repairs and allows for more sophisticated testing equipment and better quality control. Tests done at service stations are often too inaccurate to disclose a problem.

• **Frequency of testing.** More frequent testing is better than infrequent testing and is especially important for older vehicles.

• **Heavy-duty testing.** Heavy-duty trucks are major emitters of fine particulate and NO_x, but effective testing is expensive. Both BC and Ontario are implementing heavy-duty testing. The most effective programs involve centralized testing for a number of pollutants, but due to cost concerns both Ontario and BC restrict heavy-duty diesel truck tests to opacity (i.e., visible smoke). This helps reduce particulate emissions but not NO_x.

Eliminating the worst polluters: Accelerated vehicle retirement programs

Older vehicles often lack effective pollution control equipment and are a severe pollution problem. Accelerated vehicle retirement programs remove these vehicles from the road by buying the car and then scrapping it.[34] Under BC's Scrap-It program, incentives are offered by program sponsors including auto dealers. The incentives include either $750 towards the purchase of a new vehicle or a transit pass for one year. The majority of participants choose a transit pass.[35] The effectiveness of programs can be increased by:

• Targeting vehicles that are being driven further. Scrap-It only accepts vehicles that have been insured in the previous two years

The federal and many provincial governments exempt some alternative fuels from fuel sales and excise taxes.

and are still functioning. Inspection and maintenance programs often track odometer readings, potentially allowing programs to offer special incentives for vehicles that are being driven further.

• Targeting vehicles that are the grossest polluters. Scrap-It accepts vehicles that are model year 1982 or older and have failed the AirCare test. Potentially, higher bounties could be offered for vehicles that qualify for conditional passes or fail AirCare by a greater margin.

• Increasing the number of vehicles scrapped. In its first year, Scrap-It disposed of 750 cars. The smaller the program, the greater the likelihood that the vehicles being scrapped would have been scrapped anyway.

While accelerated retirement programs may remove more fuel inefficient vehicles from the road, they may not yield lifecycle reductions in greenhouse gases. This is because the programs may accelerate sales of new vehicles, which tend to be driven further and have large greenhouse gas emissions associated with their manufacture.

Conclusion

This chapter identifies a variety of ways that different players – governments, industry, and the public – can significantly contribute to the twin goals of reducing our dependence on the motor vehicle and reducing harmful air emissions from the vehicles that do travel our roads and highways.

A key theme has been the necessity of factoring air pollution concerns into the way we plan our cities, towns, and transportation systems. To do this requires a vigilant, informed, and active public.

Another key area where public involvement can and has made a difference concerns air pollution from stationary industrial sources. The next chapter addresses this topic.

	Proposed Federal Sulphur Regulation	Federal Benzene	Ontario & Quebec	BC Cleaner Gasoline Regulation	US Federal RFG Phase II	California RFG
Geographic Application	Canada-wide	Canada-wide	Southern Ontario Outaouais-Montreal Corridor	BC – variable	Most polluted areas	State-wide
Implementation Date	2005	1999	1998	Full implementation phased in 1996 to 2001	2000	1996
Maximum Summer Vapour Pressure (kP)			62	55 (Lower Fraser Valley)	47	47
Maximum Benzene (yearly average percent volume)		0.95		0.8 (province-wide)	0.8	0.8
Maximum Sulphur (yearly average ppm)	30			150 ppm in Lower Fraser Valley/ Van. Island; 220 ppm in other areas	130 (40 standard anticipated in 2005)	30
Maximum Aromatics (yearly average percent volume)				31.3 (province-wide)	25	22
Maximum Oelefins (yearly average percent volume)				13.9 (province-wide)	8.5	4

Table 1 **North American Gasoline Standards**

Source: Based on California Air Resources Board, "Fact Sheet 3: Comparison of Federal and California Reformulated Gas," April 19, 1999, and *Cleaner Gasoline Regulation*, BC Regulation 498/95. See web-site: www.qp.gov.bc.ca/stat_reg/regs/elp/r498_95.htm

Endnotes

1 Environment Canada, *State of the Environment*, 1996, at 11-89, and Environment Canada, State of the Environment Bulletin 96-5, Canadian Passenger Transportation.

2 Hydrogen is typically produced either through electrolysis or stripping hydrogen from fossil fuels. Both processes use significant amounts of energy and stripped fossil fuels produce carbon dioxide. Electrolysis requires electrical energy. In most of North America (including areas that rely mainly on hydroelectricity), increases in electricity demand are met by fossil fuel combustion causing pollution. See David Suzuki Foundation/Pembina Foundation, *Climate-Friendly Fuel* (Vancouver, March 2000).

3 At 30 people per hectare densities or less, bus systems are rarely economic while at 40 people per hectare they become economically sustainable and at 60 become completely self-sustaining. In addition, people in high-density areas drive one-third as much as people in low-density districts, and as many as one-third of households do not own a car at all. British Columbia Roundtable on the Environment and Economy, *State of Sustainability: Urban Sustainability and Containment* (Victoria: Crown Publications, 1994)

4 Durning, A., *The Car and the City* (Seattle, Wash.: Northwest Environment Watch, 1996).

5 *Ibid.*

6 See, for example, the initiative called "The National Task Force to Promote Employer Provided Tax-Exempt Transit Passes" on web-site: www.cutaactu.on.ca/advocacy.htm

7 In Greater Vancouver, the revenue generated by motor vehicle license fees and fuel taxes between 1983 and 1992 was only 80% of the cost of road building and maintenance, but gasoline is exempted from the provincial sales tax. Motor vehicle drivers in Greater Vancouver are estimated to receive an annual subsidy of $2,600 per car. Eliminating these subsidies to cars would yield a $1,145 reduction per household in provincial and municipal taxes.

8 Dedication of tolls to transportation can increase their palatability but they should be dedicated to all forms of urban transportation, not just new roads and highways. Dedicating revenues from car use to increased transit and trip-reduction programs provides revenues to improve and encourage alternatives

9 Litman, Todd, "Distance based vehicle insurance: A practical strategy for more optimal pricing" (Victoria, BC: Victoria Transport Policy Institute, 1999), available at searchpdf.adobe.com/proxies/1/11/51/58.html

10 Dedication is politically easiest when congestion is limited, but will become more difficult as car users demand the space. Creating new highway space for buses and high occupancy vehicles is more problematic. On the one hand, adding an HOV lane allows for an increase in capacity that decreases congestion, but as road capacity increases, so will the number of cars. On the other hand, HOV lanes can provide

an incentive to car-pooling or bussing. For this to be the case, occupancy requirements need to be sufficiently stringent so there is a significant time savings from car-pooling and taking transit. Generally, occupancy requirements of three or more are a minimum.

11 See news release at web-site: www.ec.gc.ca/press/00519_n_e.htm

12 *Motor Vehicle Emissions Reduction Regulations* (BC Regulation 517/95). See web-site: www.qp.gov.bc.ca/stat_reg/regs/elp/r517_95.htm

13 See *Backgrounder*, "BC Evaluates Post-2004 Motor Vehicle Emission Standards," at: www.elp.gov.bc.ca/main/newsrel/fisc9900/march/bg190a.htm

14 California Air Resources Board, "Reformulated Gas Backgrounder 6 and Fact Sheet 3: Comparison of Federal and California Reformulated Gas," April 19, 1999. See web-site: www.arb.ca.gov.

15 Mittelstaedt, M., 1998, "Auto, petroleum industries face off over cleaning Canada's gasoline," *Globe and Mail*, May 27, 1998, A3.

16 Fact sheet, "Government of Canada Actions on Clean Air" (May 19, 2000), website: www.ec.gc.ca/press/00519h_fe_htm; see also Government Working Group on Sulphur and Diesel Fuel (Ottawa: Environment Canada, July 14, 1998).

17 Press release, Sierra Legal Defence Fund, "Imperial Oil's high-sulphur gasoline triggers boycott call from Friends of the Earth" (April 19, 2000). See web-site: www.sierralegal.org

18 *Diesel Fuel Regulations* SOR/97-110. See web-site: http://canada2.justice.gc.ca/FTP/EN/regs/Chap/C/C-15.3/SOR97-110.txt

19 Initial studies indicate a 400 ppm standard would avoid 46 premature deaths, 33,700 restricted activity days, and 242,000 acute respiratory symptoms per year by 2020.

20 David Suzuki Foundation and Pembina Institute for Appropriate Development, *Canadian Solutions: Practical and Affordable Steps to Fight Climate Change*, October 1998. See web-site: www.davidsuzuki.org/PDF/CNDSolutionsNEW.pdf

21 The *Motor Vehicle Fuel Consumption Standards Act*. See web-site: http://canada.justice.gc.ca/FTP/EN/Laws/Title/M/index.html

22 The Rocky Mountain Institute Homepage is www.rmi.org/index.html

23 Pembina Institute for Appropriate Development and David Suzuki Foundation, *supra*; also, Canada, Natural Resources Canada, Energy Sector, "Model Simulations of the Climate Action Network Program for Energy Demand, GHG Emissions and Investment" (June 1995). See web-sites: www.pembina.org and www.davidsuzuki.org

24 Green, David L., "Vehicle use and fuel economy: How big is the 'rebound' effect?" *The Energy Journal* 13 (19), at 117-143; also, Litman, *supra*.

25 Green, *supra*, at 119. Cars are subject to the tax if their efficiency is below 22.5 miles per US gallon (10.6 litres per 100 kms).

26 DeCicco, John M., and Gordon, Deborah, "Steering with Process Fuel and Vehicle Taxation as Market Incentives for Higher Fuel Economy," in *Transportation and Energy: Strategies for a Sustainable Transportation System* (Berkeley, Cal.: American Council for Energy-Efficient Economy, 1995), at 183.

27 International Institute for Sustainable Development, *Making Budgets Green, Leading Practice in Taxation and Subsidy Reform* (Winnipeg, Man.: IISD, 1994), at 10-11. The rebate available to the most fuel efficient cars ($100) is a small fraction of the price of a car. The system raised $30 million in 1992.

28 Greene, *supra,* at 124.

29 Greene, *supra,* predicts increased costs of $300 per car to increase average fuel efficiency to 33 mpg by 2000.

30 British Columbia, Ministry of Environment, Lands and Parks and Ministry of Energy, Mines and Petroleum Resources, *British Columbia Greenhouse Gas Action Plan,* November 1995. See web-site: www.publications.gov.bc.ca/publications/queries/mainpagee03search.asp

31 Information on tailpipe emissions is often conflicting with some reports showing higher tailpipe emissions of NO_x from propane and natural gas and others showing lower emissions. See Clean Fuels Consulting Inc., "Comparative Analysis of Alternate Transportation Fuels," 1994 (unpublished), also US Department of Energy, *Alternatives to Traditional Transportation Fuels,* 1993 (Washington: Energy Information Administration, 1995).

32 Stewart, Steve, and Gourley, Dave, *AirCare Program Review and Evaluation of Benefits, program Years One to Five, September 1992 to August 1997.* Available at: www.aircare.ca

33 BC requires emission performance warranties but the terms are less stringent than US requirements. See *Motor Vehicle Emissions Control Warranty Regulation,* BC Regulation 116/96, available at: www.legis.gov.bc.ca/

34 See Kallen, R., "Components of a Model Accelerated Vehicle Retirement Program," *Environmental Law and Policy Center* (March 27, 2000). See web-site: www.elpc.org

35 Innovatech Energy Systems Ltd., "Evaluation of the Scrap-It Pilot Program," August 1997.

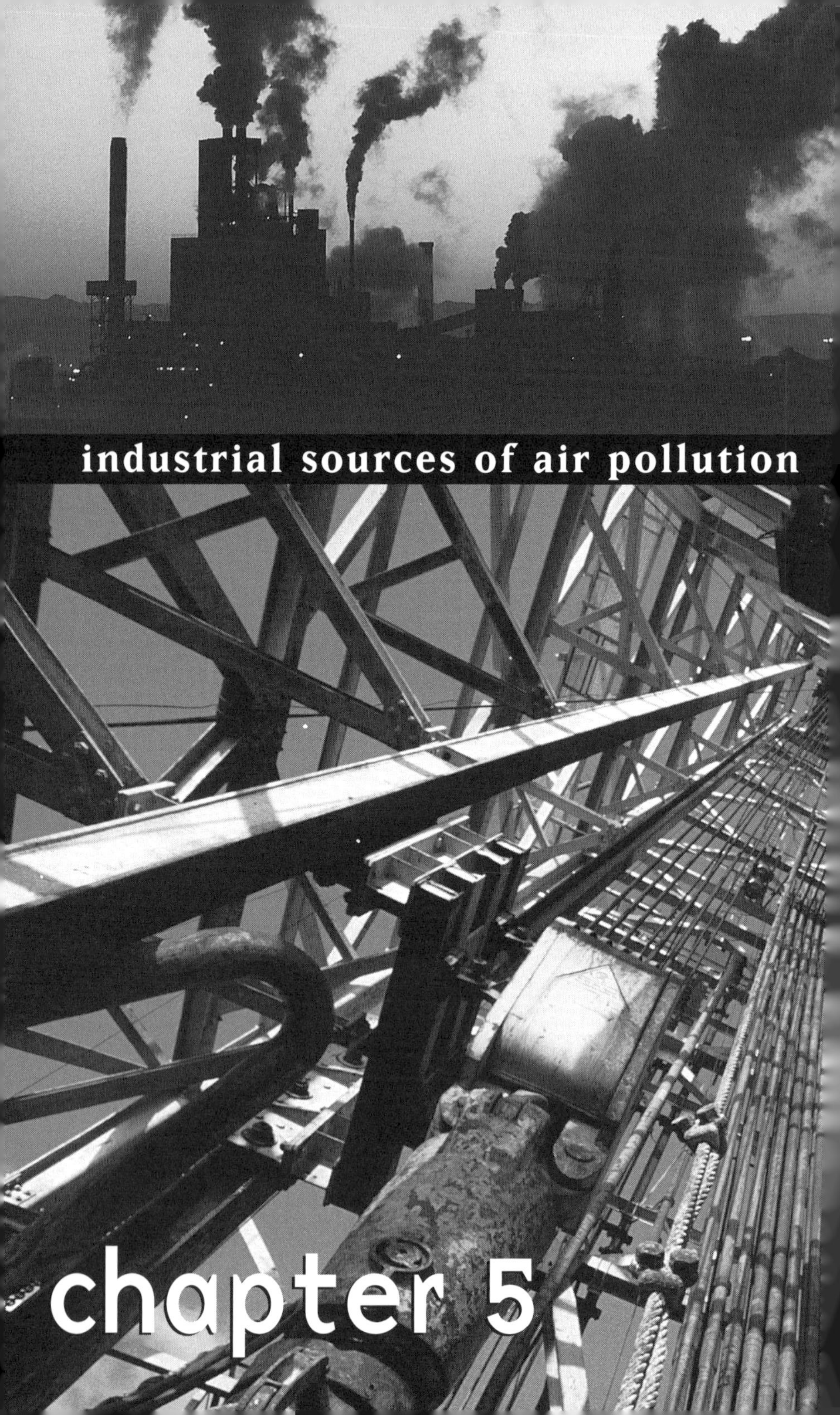

industrial sources of air pollution

chapter 5

Most industrial sources of air pollution are stationary. In other words, they are emitted by a factory or other geographically confined point.

Introduction

This chapter focuses on how governments currently attempt to regulate and manage industrial source air pollution, and how – through enhanced public participation and other means – we can collectively do a more effective job of combatting it.

Most industrial sources of air pollution are stationary. In other words, they are emitted by a factory or other geographically confined point. In general, stationary sources of many air pollutants fall under provincial jurisdiction, except where they are federally owned or run facilities, where they are located on federal lands, or where their emissions have transboundary international effects. The federal government is also empowered to regulate emissions of "toxic" substances from stationary sources under special procedures set out in the *Canadian Environmental Protection Act.*

Legally binding restrictions on industrial air pollution tend to take the form of either regulations or site-specific permits. In some jurisdictions, notably Ontario, most industrial source emitters are governed by both regulations and permits; in others, such as BC, they are primarily governed by site-specific permits. New sources of industrial emissions are also potentially subject to legal review under federal and provincial environmental assessment legislation.

As discussed in Chapter 4, governments have increasingly been exploring the use of voluntary programs of various kinds aimed at reducing emissions. This chapter considers four main approaches or strategies governments have adopted to control and reduce industrial source air pollution:

- air quality objective setting
- air emission regulation and permitting regimes
- environmental assessments of new or expanded air emission sources

- voluntary approaches including pollution prevention planning, voluntary emission reduction programs, and emissions trading

Air Quality Objective Setting

As discussed in Chapter 3, one of the key policy tools employed by governments to manage air quality are ambient air quality objectives. These objectives identify levels (or concentrations) of pollutants in the air that are deemed to be scientifically acceptable. Although these standards have no legally binding force, they are relied on by government regulators for guidance in the issuance of emission permits, in the development of regulations and policies, and in the implementation of emergency plans to deal with unacceptable air quality.

While most Canadian jurisdictions have established ambient air quality objectives, they have been criticized on a variety of grounds by environmental and public health organizations. A key criticism is that these objectives do not require or mandate government action. Another is that often these objectives were set many years, or even decades, ago and as such do not reflect current scientific knowledge with respect to the health impacts of particular pollutants. Moreover, in most jurisdictions, objectives exist for only a small handful of harmful air pollutants.

A case in point are the federal government's National Ambient Air Quality Objectives established under the *Canadian Environmental Protection Act*. To date, the federal government has only set national objectives for five substances (see table below), and they were originally set in the early 1970s.

Another criticism of ambient air quality objective-setting is that it has led to inconsistent standards and approaches being adopted across Canada as a whole.

Partially in response to these concerns, the Canadian Council of Ministers of Environment (CCME) are currently negotiating Canada-wide Standards (CWS) for priority pollutants of national concern under the auspices of the Canada-Wide Accord on Environmental Harmonization (1998). The goal of the CWS process

Governments also manage air quality by imposing legally binding requirements on industrial emitters.

National Ambient Air Quality Objectives (Canada)

Maximum Acceptable Concentration	Maximum Desirable Concentration
Ozone (O_3)* 0.082 ppm in 1 hour	0.050 ppm in 1 hour
Nitrogen dioxide (NO_2) 0.053 ppm annual	0.032 ppm annual
Carbon monoxide (CO) 31 ppm in 1 hour	13 ppm in 1 hour
Sulphur dioxide (SO_2) 0.344 ppm in 1 hour	0.172 ppm in 1 hour
Total Suspended Particulate (TSP)* 70 micrograms/m^3 annual	60 micrograms/m^3 annual

Source: Environment Canada (1998). See www.ec.gc.ca/emission/1-3e

*The objectives for ozone and TSP are currently under review as part of the Canada-wide Standards process (see below)

is to develop national standards and "to prepare complementary workplans to achieve those standards based on the unique responsibilities and legislation of each government."[1] Once agreement is reached on a CWS, implementation within a particular jurisdiction becomes the exclusive responsibility of the government of that jurisdiction. As such, each government will be allowed to decide what measures it will take to achieve compliance with the CWS. Consequently, as with current ambient air quality objectives, citizens will not have legal recourse to require their government to take particular steps to implement a CWS, although governments will be under a new obligation to report on their progress towards attaining the agreed-on standard.

In November 1999, draft CWS for four substances were presented to the CCME for consideration: fine particulate matter, ground-level ozone, benzene (phase 1), and mercury from incineration and metal smelting. The CCME has accepted in principle standards for PM and ozone that are a significant improvement over comparable National Ambient Air Quality Objectives. Final standards are expected by the summer of 2000. Proposed standards are also in the works for mercury emissions from electrical power generation (spring 2000) and benzene in air (phase 2) (spring 2001).[2]

Air Emission Regulation and Permitting Regimes

Governments also manage air quality by imposing legally binding requirements on industrial emitters. These requirements can take the form of general prohibitions against certain types or amounts of emissions (see discussion of regulations in Chapter 3), or they can be set out in a permit issued to a particular industrial facility that confers the right to emit designated air contaminants according to the terms of that permit.

If an industrial source emitter violates a regulation applicable to its activities or fails to comply with the terms of its emission permit, government is entitled to take enforcement action including prosecution. If government fails to do so, a member of the public can commence a private prosecution (see Chapter 6). Members of the public are also usually entitled to comment on new regulations before they are brought into force and on permits before they are issued. In the latter case, the public is also entitled to appeal a decision to grant or renew a permit (see Chapter 6).

In this section of the chapter, we will focus on the regulation and permitting regimes in place at the federal level and in Ontario and British Columbia. For information on other Canadian jurisdictions, see the Appendix.

Air emission regulations

Regulations may stipulate emission limits that must be incorporated into permits for specific sectors or facilities. Alternatively, regulations may be used to control specific products or standards,

The federal government is empowered to regulate a substance once it has determined that it is toxic and placed it on an official List of Toxic Substances.

such as ozone depleting substances. Regulations controlling specific industrial sources of air pollution are found mostly at the provincial level. Federal regulations focus primarily on toxic substances.

Federal regulation

At the federal level, the primary law governing pollution is the *Canadian Environmental Protection Act* (*CEPA*).[3] *CEPA* empowers the federal government to control pollutants in two ways. The first method is to prohibit the release of substances classified as "toxic" at levels dangerous to the environment or to human life or health. *CEPA* also empowers the federal government to regulate if an emission affects federal lands, emanates from a federal government operation, or has international transboundary effects.

The federal government is empowered to regulate a substance once it has determined that it is toxic and placed it on an official List of Toxic Substances. Regulation of substances on the list is intended to be comprehensive, "from the cradle to the grave" – i.e., from a substance's production and transport to its use and emission. However, the regulation of toxics has focused on substance emissions, ignoring substance production, transport, and use.

CEPA's limitations include the broad ministerial discretion to designate substances as toxic and to recommend regulatory action.[4] As a result of this discretion, very few of the 30,000 to 40,000 chemicals estimated to be manufactured or imported into Canada are regulated. The only listed toxic substances emitted into the air that are currently regulated under *CEPA* are asbestos, mercury, lead, vinyl chloride, PCBs, and ozone depleting substances.

Federal government involvement in air issues could expand significantly if the federal government follows through on its announced intention to list PM10 as a toxic substance. The Ministers of Environment and Health have recommended that PM10 be added to the List of Toxic Substances.[5] Because the sources of PM pollution are ubiquitous, this could potentially lead to a far greater role for the federal government in regulating air emissions.

As noted above, *CEPA* also allows the federal government to control air emissions of non-listed substances where the emission has adverse international transboundary effects. To date, however, the federal government has never deployed this regulatory power.

Finally, *CEPA* allows the federal government to regulate fuel standards, a power that has been used to restrict sulphur content in gasoline and diesel.

Ontario

The *Air Pollution Regs.* are the main vehicle by which the Ontario government regulates air quality. This regulation sets maximum concentration limits for close to 100 air pollutants at "points of impingement," or hypothetical locations where contaminants are expected to contact the ground.[6] It is illegal to release a contaminant at levels that result in a point of impingement standard being exceeded.[7] Point of impingement standards represent minimum levels of protection – the government can impose stricter requirements on specific industries through permitting requirements.

Other regulations passed under the *Environmental Protection Act* control air emissions from specific sources, including waste incinerators, ferrous foundries, hot mix asphalt facilities, and mobile PCB destruction facilities. Ozone depleting substances are also controlled by regulation in Ontario, as are sulphur dioxide emissions.

In 1999, the Ontario government announced new air emission caps for the electricity sector, which is responsible for a significant portion of emissions of local smog precursors. These caps require Ontario Power Generation (Ontario's primary electricity generation utility and formerly Ontario Hydro) to reduce its emissions of NO_X and SO_2 by 5 and 10% respectively. However, these caps will allow substantially higher emissions than those recommended by the Ontario Medical Association to ensure public health. In addition, eight key power plant emission pollutants were left uncapped, including mercury, carbon dioxide, and lead.[8]

Provincial governments can also empower municipal governments within their jurisdiction to enforce provincial air pollution laws or regulations, or to pass their own air pollution bylaws.

British Columbia

In BC, the *Waste Management Act* is the primary statute under which industrial sources of air pollution are controlled. The Act generally prohibits the release of waste (which includes air contaminants); however, pollutants may be emitted into the air if statutory exclusions, regulations, or permits allow for their release.

Only a few regulations relate specifically to industrial source air pollution. Regulations have been used to control open burning of debris,[9] as well as hazardous waste incineration[10] and ozone depleting substances. Regulations have also been used to phase out wood residue (beehive) burners and replace them with cleaner incinerators or alternative technology.[11]

Unlike Ontario, BC tends to regulate industrial source air pollution on a source-specific basis (with the exception of the regulations listed above). This approach has some advantages. For example, where industrial emissions in one location have a greater cumulative environmental impact than emissions in another, each source can be treated differently. However, the downside is that it can lead to a lack of uniformity in standards across the province.

Local regulation

Provincial governments can also empower municipal governments within their jurisdiction to enforce provincial air pollution laws or regulations, or to pass their own air pollution bylaws. For example, in BC, the *Waste Management Act* gives the Greater Vancouver Regional District (GVRD) the power to pass bylaws to control or prevent air polluting emissions from sources within the GVRD. The GVRD has enacted two bylaws in an attempt to control the contaminants from point source pollution, using a permitting system similar to that of the province.[12]

Municipalities can also control industrial air pollution by relying on their powers to pass licensing and nuisance bylaws. Nuisance bylaws ensure operational facilities do not produce noxious air emissions. Bylaws may also limit hours of operation, thereby reducing emissions. [13]

Permitting regimes

Environmental protection legislation in all provinces establishes permitting regimes for new sources of pollution. Permits are, by and large, the most common tool used to control industrial source air pollution in Canada.

All provinces require that point sources of air pollution (e.g., factories, mills, electricity generation plants) obtain permits or approvals before any new or modified source of pollution is put into operation. These approvals generally contain emission limits, monitoring and reporting requirements, and review or expiry provisions.

Emission limits that are prescribed in permits often reflect ambient air quality objectives established by either the federal or provincial governments. However, the incorporation of these objectives is at the discretion of the permitting agency or air quality manager. Regulatory standards, on the other hand (such as Ontario's point of impingement standards), must be incorporated into permits.

Ontario

Permits under the Ontario *Environmental Protection Act* are called "certificates of approval" (or C of As). Before constructing or altering equipment or production processes that discharge pollutants, industrial sources must obtain a C of A from a government official (known as the Director). The Director must not allow the project to proceed if resulting emissions would exceed point of impingement concentration levels set out under the *Air Pollution Regs.*, emission limits set out in other regulations, or if an "adverse effect" would result. The Director has discretion to demand stricter standards than those outlined in regulations if an adverse effect could result from a certificate being granted. However, in practice, this rarely occurs.

Taylor, BC

Citizens have the right to appeal permitting decisions (see Chapter 6). For example, in 1996, Sierra Legal Defence Fund successfully appealed a decision to issue a permit for a hazardous waste incinerator. Acting for a small citizens group based in Taylor, BC, SLDF argued that the permit did not contain conditions that adequately addressed the health concerns of local residents. The Appeal Board agreed, ordering that rigorous new requirements be imposed on the operation under its permit.

British Columbia

In order to release air contaminants in British Columbia, industrial point source emitters must apply for permits, which are granted by regional managers of the BC Ministry of Environment, Lands and Parks. Managers may consider the BC government ambient air quality objectives (known as "Pollution Control Objectives") as well as other sources of information in deciding whether to grant a permit and in establishing permit conditions.[14]

Because of the broad discretion exercised by managers, permitting standards imposed on a facility in one region may differ from those imposed in another. In addition, given the number of permits issued and the discretion involved in their issuance, it is often difficult for the public to become involved or track permitting decisions.

Environmental Assessment

Legislated environmental assessment is a relatively new phenomenon in this country. Until recently, in many Canadian jurisdictions there was no legal requirement that major new projects with potentially significant environmental impacts undergo a formal environmental assessment. Now, when the public has concerns about the impacts of a proposed new project, development, or activity in terms of air quality, it can urge that an environmental assessment be conducted.

The purpose of environmental assessment is to ensure early consideration of potential environmental impacts from new industrial activities or facilities. Environmental assessment identifies costs in terms of environmental degradation, use of energy and resources, and social and economic disruption. Environmental assessment helps decision-makers balance these costs against known benefits, such as employment, revenue, and the necessity of the product or service.[15]

Environmental assessment legislation is in place both federally and provincially. The level of government responsible for an assessment depends on a number of factors, including constitutional authority. In general, provinces assess private business and provincial government projects and activities. Federal assessments usually address projects on federal land or those conducted by federal government agencies, as well as projects of broader national or public concern (see below).

The following discussion is provided to give a sense of the circumstances in which environmental assessments are conducted at the federal and provincial levels, as well as the nature and scope of the assessment that is likely to take place. For further information on environmental assessment sources, see Appendix.

Canadian Environmental Assessment Act (CEAA)

Under *CEAA*, environmental assessments are required for all projects where the federal government:

- is the party (or proponent) proposing it
- provides financial assistance
- issues permits or approvals under federal legislation or regulations in order to enable the project (these permits and approvals are listed in the *CEAA* Law List Regulation)
- transfers an interest (e.g., sells, leases, or grants a right of way) in federal lands (including Indian reserves) in order to enable the project.

The Minister of Environment can also refer a project to federal environmental assessment when he or she is of the opinion that it

Over the last decade or so, considerable attention has been paid to identifying ways to create incentives for environmental protection and enhancement.

may have significant adverse impacts in a province other than the one in which it is located, internationally, or on federal lands. As such, if a proposed project is likely to have transboundary impacts on air quality, the public would have a sound basis to lobby for a federal environmental assessment.

Three types of assessment are possible under the *CEAA*: a basic screening; a comprehensive study; or a full-scale public hearing. The level of assessment is determined on the basis of both the size and anticipated production levels of the project or activity. Opportunities for public input into an environmental assessment are mandatory for comprehensive studies. Intervenor funding for public interest groups is available for public hearings only.

Only a very small number of the federal permits that trigger assessments (such as for mobile PCB destruction facilities) relate directly to air quality. As a result, federal environmental assessment has not been used regularly to directly address air pollution. However, assessments triggered by an application for non-air-related permits can, and often do, entail consideration of air impacts.

Ontario *Environmental Assessment Act*

Ontario's *Environmental Assessment Act* applies only to provincial or municipal developments or "undertakings." Private sector projects or activities are not subject to the Act unless they receive special designation by the Minister of the Environment. In practice, this has rarely occurred.

The environmental assessment process requires a project proponent to submit an environmental assessment document describing the project to the Minister of the Environment. Proposals are also made available to the public via the Environmental Registry.

After the government and the public have had the opportunity to review the document, anyone can request the Minister of the Environment to refer the proposal to the Environmental Assessment Board (an independent, quasi-judicial body) for a hearing. If the application is referred, the board has the power to accept or amend the assessment document, approve the undertaking as presented, or approve the undertaking subject to specified terms and conditions. It should be noted, however, as is typical in environmental assessment processes, that assessment board recommendations are not binding on the elected officials who have the ultimate power to decide whether a project should proceed.

British Columbia Environmental Assessment Act (BCEAA)

In BC, all projects and activities described in the *Reviewable Projects Regulation* are automatically subject to an environmental assessment under *BCEAA*. Assessments are generally triggered by the construction of major new projects or major modifications to existing operations.

Proponents of reviewable projects must apply for an approval certificate. Applications must be reviewed by a project committee, composed of representatives from government, First Nations, and neighbouring jurisdictions. Upon review, the project committee can either forward the application to the responsible Cabinet Minister with a recommendation that a permit be granted or require the preparation of an in-depth environmental assessment (or project report). Project reports must include both detailed information on potential environmental impacts, as well as any means to incorporate energy efficiency and conservation measures (or otherwise reduce negative impacts) into the design and operation of the project.

In exceptional circumstances, government may refer to the Environmental Assessment Board for public hearings. Although public hearings are unusual, the Act contains a variety of public notice and comment provisions and applications can be tracked through an online registry (see Appendix).

> Pollution prevention is intended to apply to every stage of industrial production: extraction, manufacturing, consumption, and disposal.

Voluntary Approaches

Over the last decade or so, considerable attention has been paid to identifying ways to create incentives for environmental protection and enhancement. Business has mounted a concerted lobby to persuade government to replace mandatory, prescriptive regulation with more voluntary, market-based policies and approaches. Environmentalists initially were sceptical of the ability of the market to deliver environmentally optimal results. Of late, however, many have decided that carefully defined programs can usefully supplement, although never entirely replace, effective regulation designed to protect "bottom line" environmental and health concerns.

In this final section, three emerging voluntary approaches to environmental protection that are becoming increasingly important in the area of industrial source air pollution are addressed:

- pollution prevention planning
- voluntary emission reduction programs
- emissions trading programs.

Pollution prevention planning

Traditionally, Canadian governments have adopted a "pollution control" approach as a primary means of protecting the environment. An illustration of this are laws that require industries to fit exhaust stacks with scrubbers or other pollution control devices to clean emissions. However, these devices often do no more than disperse pollution (by emitting pollutants farther away) or take pollutants from one place (e.g., air) and put them in another (e.g., water).

In contrast to these traditional "end of pipe" techniques, pollution prevention seeks to control the *creation* of pollutants. Pollution prevention is intended to apply to every stage of industrial production: extraction, manufacturing, consumption, and disposal.

Pollution prevention techniques have many advantages over traditional approaches. These include:

- conservation of materials, therefore slowing depletion of virgin resources and reducing energy consumption
- reduction of costs to industry by minimizing handling, storage, and treatment of wastes
- reduced costs to society by minimizing the need to clean up toxic problems.

Pollution prevention techniques are generally implemented by plans that consider the successive stages in the lifecycle of products, with the goal of avoiding, eliminating, or reducing pollution. Effective plans protect human health (including workers, consumers, and the general public) and the environment. In addition, they optimize economic opportunities and reduce economic risk, operating costs, and corporate risk of criminal and civil liability.[16]

Pollution prevention planning has been largely voluntary in Canada in recent years. However, federal and provincial governments have started to empower their ministers to require pollution prevention plans from industry.

For example, the 1999 amendments to the *Canadian Environmental Protection Act* establish pollution prevention as a national goal. To this end the Act as amended allows the Minister of the Environment to require pollution prevention plans for toxic substances designated under the Act. With Cabinet approval, pollution plans can also be required for sources of international air pollution.

Companies are also increasingly aware of the competitive advantages of adopting comprehensive pollution prevention plans as part of a broader environmental management system (EMS). Once a company has adopted and implemented an EMS, it can apply to be certified by the International Standards Organization under the ISO 14001 program. ISO certification can open new market opportunities for a company.

Voluntary emission reduction programs have become increasingly popular in Canada, particularly with respect to air contaminants.

Voluntary emission reduction programs

Voluntary emission reduction programs have become increasingly popular in Canada, particularly with respect to air contaminants. Some of these programs are led by industry, such as the Responsible Care program, which was developed and implemented by the Canadian Chemical Manufacturers Association. Other programs are initiated by governments, such as the Ontario Smog Plan. Governments also use codes of conduct and agreements with industry sectors to encourage voluntary reductions of a variety of pollutants.

One of the best known voluntary programs for industrial sources of pollutants is the federal Accelerated Reduction/Elimination of Toxics (ARET) program. Created in 1994, ARET calls on industry to voluntarily reduce its emissions of twenty-nine persistent toxic substances by 90% by the year 2000, and achieve a 50% reduction of other substances covered by the program.

Proponents claim ARET has already been a significant success. Participating facilities reduced their emissions of toxic substances by almost 21,500 tonnes within two years – a decrease of over 60%. However, the 300 participating facilities represent less than 20% of the facilities required to report their annual emissions to Environment Canada (a requirement based on emission levels of these pollutants).[17]

ARET and other similar programs have therefore been criticized by environmental groups for not achieving real improvements to air quality and for being virtually impossible to monitor. These criticisms have been echoed by independent watchdogs, including the federal Commissioner for Environment and Sustainable Development. In 1999, the Commissioner reported to Parliament that neither ARET nor other federal voluntary agreements "are sufficient to be used as the only tool for achieving and measuring reductions of priority toxic substances."[18] The Commissioner also raised concerns about the lack of independent evaluation and monitoring of these agreements.

Emissions trading

Emissions trading has become a source of increasing interest in government and industry circles. While emissions trading is prevalent in the United States, it is only now beginning to establish a presence in Canada. In the United States the most commonly used emissions trading regime is the so-called "cap-and-trade."

Under a cap-and-trade system, governments identify key emission sources, set a limit or cap on overall emissions, and allocate a portion of that cap to each source. Each source has three options: it can reduce its emissions to meet its allocation, it can reduce emissions below its allocation and sell surplus reductions as "allowances," or it can purchase surplus allowances from other sources. The theory is that inefficient, high-polluting companies will develop more efficient, clean technologies or ultimately be forced out of business by the high cost of purchasing credits. Although actual trading is voluntary, the US cap-and-trade programs are based in law, with strict rules on trading and emissions, and penalties for non-compliance.

In contrast, emissions trading in Canada to date has been purely voluntary and involved "credits" trades. Credits are generated by any real reductions below regulated limits; so far, however, overall caps on emissions have not been imposed. For example, a Pilot Emission Reduction Trading program (PERT) was initiated in the Windsor-Quebec corridor in 1996. PERT is a multi-stakeholder process involving government departments, Ontario Power Generation (OPG), and health and environment groups. It focuses on reducing smog and other pollutants by allowing trades in nitrogen oxide, volatile organic compounds, and greenhouse gas emissions. Trades do not currently have any legislative or regulatory force, although there is an expectation among participants that PERT credits will be recognized under any future binding programs.

In British Columbia, a Greenhouse Gas Emission Reduction Trading pilot (GERT), is under way. Under this program, credits are generated for reductions in greenhouse gases. Participants in the program have been given a commitment that governments will rec-

Emissions trading has become a source of increasing interest in government and industry circles.

ognize the credits in the context of future regulatory programs.

The most recent and controversial initiative in this area has been an Ontario government plan that will allow OPG to meet newly imposed emission caps on acid gas (discussed earlier in this chapter) by purchasing emission credits from Canadian and US companies located in Ontario's airshed. This approach has been strongly criticized because it is not a true cap-and-trade system. The only company that is capped under the program is OPG. Other companies will be allowed to earn emission credits (by instituting pollution reduction technologies that reduce emissions from their operations), sell these credits to OPG, and then proceed to expand their own production with the result that they end up returning to their original emission levels.[19]

As a result, critics predict that the plan might mean that net emissions into Ontario's airshed will remain constant or possibly increase. For emissions trading to have a beneficial effect on net emissions, companies should only be able to trade credits that reflect the actual amount by which they have reduced emissions.

Conclusion

Governments and businesses have been paying a lot more attention to industrial source air pollution in recent years. Yet there is much to be done: standards are out of date; lists of toxic substances are incomplete; there are serious gaps in our scientific knowledge, especially as to the synergetic effects of multiple pollutants; and source monitoring and enforcement is often inadequate.

Many governments and most businesses generally continue to favour voluntary approaches that seek to achieve change through incentives and moral suasion. While many environmentalists and public health advocates support and are involved in some of these initiatives, there is a growing impatience to see results. As a result, we are increasingly seeing activists exploring creative new means to tangibly advance the clean air agenda, including legal action. In the next and final chapter, we identify and discuss various ways citizens can better equip themselves with legal and other tools to take on this challenge.

Endnotes

1 "Canada-wide Standards: Overview," www.ccme.ca/ccme

2 *Ibid.*

3 To view *CEPA* see http://canada.justice.gc.ca/FTP/EN/Laws/Chap/C/C-15.3. For information on recent *CEPA* amendments, see web-site: www.ec.gc.ca/CEPA/index_e.html

4 See Vanderzwaag, D., and Duncan, L., 1992, "Canada and Environmental Protection," in Boardman, J. (ed), *Canadian Environmental Policy: Ecosystems, Politics, and Process* (Oxford, Eng.: Oxford University Press), at 9.

5 *Canada Gazette* Part 1, May 27, 2000, Vol. 134, No. 22, at 1643

6 See Office of the Provincial Auditor of Ontario, "1996 Annual Report," Chapter 3.09 – Ministry of Environment and Energy. See web-site: www.gov.on.ca/opa/en/963-9.htm (a mathematical formula is used to determine the point of impingement).

7 *Air Pollution Regs.*, schedule 1. See web-site: 209.195.107.57

8 Ontario Clean Air Alliance (J. Gibbons), "Pollution loopholes: An assessment of Ontario's approach to air-pollution control in the electricity sector." See web-site: www.cleanair.web.net

9 *Open Burning Smoke Control Regulation*, BC Reg 145/93. An exemption is granted only if the requirements set out in the regulation are met, including the requirement to follow the Open Burning Smoke Control Code of Practice. See the Regulation at: www.env.gov.bc.ca/epd/cpr/regs/obscreg.html

10 *Special Waste Regulation*, BC Reg 63/88. Specifies requirements for the siting, construction, operation, performance, management, maintenance, and closure of facilities for the storage, use, treatment, and disposal of special waste. These controls regulate the incineration of PCB, dioxin, oil, asbestos, and other "special wastes" and requirements to continuously measure CO, CO_2, and oxygen in exhaust gas. See the Regulation at: www.env.gov.bc.ca/epd/cpr/regs/swreg.html

11 *Wood Residue Burner and Incinerator Regulation*, BC Reg 519/95. Prohibits the disposal of wood residue in a beehive or unmodified silo burner except in circumstances set out in the regulation. See the *Regulation* at: www.env.gov.bc.ca/epd/cpr/regs/wrbreg.html

12 *Air Pollution Control Bylaw* 603 ("APC Bylaw") and *Air Quality Management Bylaw* 725 ("AQM Bylaw"). See web-site: www.gvrd.bc.ca

13 See Estrin, D., and Swaigen, J., 1993, *Environment on Trial: A Guide to Ontario Environmental Law and Policy*, 3rd ed. (Toronto, Ont.: Emond Montgomery Publications, 1993), at 62-67.

14 See the Ministry of Environment, Lands and Parks web-site at: www.env.gov.bc.ca/epd/cpr/objectiv/objectiv.html and www.elp.gov.bc.ca:80/epd/epdpa/ar/aq.html, for more information.

15 Estrin, D., *supra*, at 187-191.

16 See Environment Canada, "Pollution Prevention Legislative Task Force, Final Report," 1993 at 27. For a copy, write to Inquiries Centre, Environment Canada, Ottawa, ON, K1A 0H3.

17 A total of 1,818 Canadian facilities filed reports with the National Pollution Release Inventory (NPRI) in 1996, according to the NPRI Summary Report 1996. See web-site: www.ec.gc.ca/pdb/npri/index.html

18 Report of the Commissioner of the Environment and Sustainable Development to the House of Commons 1999, Chapter 3: Managing Toxic Substances, May 1999. See web-site: www.dfait-maeci.gc.ca/sustain/Links/related-e.asp

19 Ontario Clean Air Alliance, *supra*.

a citizen's toolkit

chapter 6

Being well informed is key to effective citizen involvement in environmental decision-making.

Read enough about air pollution? Want to do something about it? This chapter is written for the activist in all of us. It discusses various ways in which you can take action on air pollution. These include:

- staying informed
- community organizing
- participating in decision-making
- pursuing legal remedies.

Staying Informed

Being well informed is key to effective citizen involvement in environmental decision-making. Fortunately, government decision-making processes are becoming more open and transparent; as a result, citizens in most Canadian jurisdictions have a number of information-gathering tools at their disposal. As noted in Chapter 1, this is part of a broader trend towards recognition of citizens' right to know, particularly with respect to pollution sources that affect their community.

Inventories

Emission inventories provide citizens with information on specific sources of air pollution. There are two national pollutant inventories in Canada. Many provinces also compile and publish similar inventories.

The National Pollutant Release Inventory (NPRI) is a database that contains information on the release of many toxic substances into air, water, and land. The federal government is required by law to provide this information to the public under the *Canadian Environmental Protection Act.* Environment Canada collects information from individual facilities across the country and publishes the results on the NPRI web-site.[1] Companies may, however, request that information provided to the federal government be kept confidential.

In addition, the federal government publishes the Criteria Contaminants Inventory every five years. This inventory provides information on emissions of common air pollutants such as sulphur dioxide and nitrogen oxides. Information contained here is collected by provinces on a voluntary basis and compiled nationally by Environment Canada.

Records of Spills and Leaks

In addition to keeping track of routine ongoing emissions, in most Canadian jurisdictions there are laws requiring the reporting of spills and leaks of chemicals and other harmful substances. Ordinarily this is treated as public information, accessible on request to the relevant ministry or department of environment.

Reports

Most governments report to the public on significant air quality and human health trends. For example, Environment Canada regularly publishes statistics related to urban air quality, including average levels of air pollutants, the number of hours annually during which ground-level ozone exceeds national objectives, and levels of airborne particulates in Canadian cities. All governments also publish progress reports on various issues or programs. Many of these reports are publicly available, primarily through departmental web-sites (see Appendix).

Some agencies also compile statistics and reports on compliance with regulatory standards. For instance, the BC government publishes a twice-yearly list of companies and municipalities with poor compliance records under the *Waste Management Act*. The Greater Vancouver Regional District publishes Air Quality Non-Compliance Lists semi-annually, to notify the public of incidents of non-compliance. Environment Canada also reports statistics on enforcement as part of its annual report on activities under the *Canadian Environmental Protection Act*.

In recent years, many Canadian jurisdictions have passed freedom of information (FOI) laws to ensure that governments don't keep public documents secret just because they might be embarrassing or contain sensitive information.

Registries

Environmental registries are databases of government information pertaining to the environment. Members of the public can use these registries to locate information on permit applications, government legislation, regulations and policy, as well as appeals and current litigation.

Ontario is the only Canadian province with a comprehensive registry.[2] Ontario's Environmental Registry provides citizens with notice of proposed activities that may affect the environment. These activities include policies, acts, regulations, and instruments (such as certificates of approval). The Registry also posts current environmental court actions. The following table gives an example of information stored on the Ontario Environmental Registry.[3]

Application for a certificate of approval as posted on the Environmental Registry.

Instrument type	Application for certificate of approval for discharge into air.
Proponent	Allied Signal Canada Incorporated. Location, City of Stratford.
Description	"This application is for a certificate of approval (air) for the installation of a controlled pyrolysis cleaning furnace to clean cured paint residues, plastic residues, grease or other organic material from metal jigs, hangers, hooks, screws and other metal parts."
Government contact person	Listed

While other provinces do not have as comprehensive registries as Ontario, information is often available on specific subjects, such as permits and pending environmental assessments (see Appendix for a complete list of provincial registries).

Freedom of information

Even when governments don't want to hand over information, sometimes they can be required to do so. In recent years, many Canadian jurisdictions have passed freedom of information (FOI) laws to ensure that governments don't keep public documents secret just because they might be embarrassing or contain sensitive information. FOI laws are generally "user-friendly," and can be a useful way to figure out whether governments are properly monitoring and enforcing their laws.

However, our FOI laws are far from perfect. In some cases, it can take a long time to secure requested information. Another problem is that governments have the right to charge the party requesting information for the cost of responding to their request. Sometimes these costs can run into the thousands of dollars. Where government wants to levy such charges, the party seeking the information can seek a fee waiver from the independent commissioner responsible for administering the applicable FOI law. In most jurisdictions, the law allows the FOI commissioner to grant a fee waiver if payment would create financial hardship and release of the information would benefit public health or safety.[4]

Occasionally, companies that have provided information to government will go to court to prevent its release. A case in point is a recent battle fought the by Sierra Legal Defence Fund to obtain information provided to Environment Canada by six major oil companies with respect to the sulphur content in gasoline being produced by their refineries (see Chapter 4).

In other cases, government ministries may turn down requests for information. When this occurs, the party making the request can appeal the decision to the FOI commissioner charged with administering the legislation in question. In BC, the West Coast Environmental Law Association routinely advises environmental groups on such appeals.

Organizing with other like-minded individuals in your community can be a powerful means to improve air quality at a local level.

Further information on FOI laws in your jurisdiction is provided in the Appendix.

Community Organizing

There is no substitute for getting organized. Community organizing can include starting your own community group, community-based air quality monitoring programs, and attracting media attention to the issues you are concerned about.

Organizing locally

Organizing with other like-minded individuals in your community can be a powerful means to improve air quality at a local level. There are many resources that can help you start and run an effective local citizens group. These include:

- ***Community Environmental Projects: From Needs Assessment to Evaluation.*** This booklet, prepared by Environment Canada (under the former Action 21 program), is an up-to-date and useful guide for community group involvement in environmental stewardship. The booklet outlines steps required to set objectives, work with others, communicate to the public, and evaluate the impact of your group's efforts. It also provides a list of useful resources.[5]

- ***Not for Profit, You Say! An Operations Manual for Canadian Non-Profit Organizations***, by Rosemary Gahlinger-Beaune.[6] This useful guide explains how to set up the organization, recruit and train board members and volunteers, run meetings, obtain funding, and maintain good publicity.

There are also a variety of sources that give detailed information on obtaining funding for your community group. One of the newer funding sources is Environment Canada's Community Animation Program, which assisted in the funding of this book.[7]

Community-based air quality monitoring

Citizens groups may wish to pursue a "hands-on" approach to improving local air quality. For example, inspections and monitoring can allow communities to take action by identifying and monitoring potential problems and preventing them from growing.

Neighbourhood inspections involve gathering appropriate information on local emissions of air pollutants. In some circumstances, it may be appropriate to identify experts to conduct studies of local facilities in order to determine air emission content and quantity. This information can then be used to work with appropriate company or government officials to effect changes.

Media

Notifying the media about the outcome of neighbourhood inspections, or about local air quality improvement initiatives more generally, can be a powerful tool in pursuing environmental goals. Sympathetic media coverage can garner public support for community-based activities and goals.

In order to obtain good media coverage, community groups must express their goals and strategies clearly and concisely. To do this, organizations need to have a well-defined central message and should present information about problems and potential solutions in the form of press releases and background documents.

One useful guide to the media is West Coast Environmental Law's *Making the News: A Guide to Using the Media.* This book explains the importance of the media, how news is made, and how to use the media to accomplish your goals – all from a Canadian perspective. It also outlines how to prepare press releases and handle press conferences and interviews.[8]

Participating in Decision-Making

Increasingly, citizens are taking advantage of opportunities to participate directly in decision-making processes that affect air quality. These opportunities exist at the local, provincial, and national levels.

Many municipalities and provinces have local air quality advi-

How to Run a Monitoring Program

Extensive background materials on how to set up and run a community air quality monitoring program have been prepared by the Alberta Environmental Law Centre. Using funds made available from a creative sentence imposed on a polluter, the Centre has developed a Community Action on Air Quality package. The package can be downloaded free of charge at www.elc.ab.ca/.

sory committees that include representatives from community organizations concerned about air pollution.

Members of the public normally have the right to comment on individual pollution permits or certificates of approval (see Chapter 5). A few provinces have formal public comment periods that precede the issuance of air emission permits. In addition (as will be discussed in the next section), citizens in many jurisdictions are entitled to appeal decisions to issue such permits.

There are also opportunities for citizen involvement in the regulatory process. In many jurisdictions, before a regulation is legally binding it must be "published" in a government reference source known as a gazette. For example, regulations proposed under the authority of any federal act must be published in the *Canada Gazette*.[9] The public has sixty days to comment on or object to regulation once it is gazetted. Copies of federal and provincial gazettes can be found in major public libraries and university law libraries (see Appendix).

Finally, members of the public can also forward their comments on proposed regulations and laws directly to their elected representatives. An up-to-date list of all federal members of parliament, senators, and members of provincial legislative assemblies can be found on federal and provincial governments web-sites (see Appendix).

Pursuing Legal Remedies

The 1990s saw a steady increase in the frequency with which citizens and citizens groups have employed legal action to protect the environment. Because legal action is costly and time consuming, it should generally be regarded as a last resort to be used only when government or industry cannot be persuaded through other means to enforce or comply with the law. In these circumstances, regardless of the ultimate outcome of the suit, commencing an action can be of significant benefit by publicly "spotlighting" the environmental risk or harm in question.

To find out if you have a case that is worth pursuing, you will need to consult a lawyer. Often there are lawyers in your community who will be prepared to take on public interest environmental cases on a reduced-fee or no-fee basis (*pro bono*). There are also law firms that specialize in providing *pro bono* public interest environmental advice and litigation services. These include the Sierra Legal Defence Fund, the West Coast Environmental Law Association, and the Canadian Environmental Law Association. As well, the Canadian Environmental Defence Fund raises funds to assist public interest litigants to retain private counsel and scientific experts. For further information see the Appendix.

Even if you are fortunate enough to secure a *pro bono* lawyer, as the client you will often be required to pay litigation-related expenses including filing fees, photocopying, and expert fees. As well, as client you may be liable to pay the costs (legal and court expenses) of the other side in the event that the suit is unsuccessful. Courts have discretion to waive the imposition of such costs on public interest litigants; nonetheless, liability for such costs is a risk that should not be ignored.[10]

There are four main types of legal action available to citizens and citizen groups to protect the environment: *statutory appeals and rights of action; judicial review; private prosecutions*; and *common law tort actions.*

Statutory appeals and rights of action

Where the goal is to challenge a decision to issue an emission, development, or siting permit, the best remedy is to pursue a statutory appeal of the decision. The availability and procedural particulars of this remedy will be described in the statute under which the permit has been issued. An initial concern will be whether you or your group has standing to pursue such an appeal. Getting standing means that a court has given permission for you to bring and argue

Success in Reducing Emissions!

One of the best known recent illustrations of a successful appeal of a government air emission permitting decision is a legal victory secured by the Sierra Legal Defence Fund and Greenpeace in 1996. In this case, the groups appealed a permit issued to Petro-Canada for a major expansion of its Mississauga refinery. After a protracted legal battle involving nineteen days of hearings and considerable expert testimony in front of the Ontario Environmental Appeal Board, Petro-Canada backed down. Under the terms of a settlement, Petro-Canada agreed to reduce its sulphur dioxide by 80% and its nitrogen oxide emissions by half. It also agreed to contribute $250,000 to a trust fund for airshed research to be administered by Greenpeace.[11]

a matter before it. Standing requirements vary from jurisdiction to jurisdiction. In some, you will need to show that you or your group will be "directly affected" by the decision to issue the permit. Other jurisdictions have less restrictive standing requirements that depend on whether you can demonstrate that you have a legitimate interest in the issues surrounding the permitted activity. If you or your group are active around air quality issues and have participated in the public process that led up to the decision to issue the permit, this might help secure standing for appeal purposes.

In some jurisdictions there are additional statutory rights that provide legal recourse to groups concerned with emission permitting or air quality enforcement issues. In Ontario, the Yukon, and New Brunswick, under the new *Canadian Environmental Protection Act*, citizens have the right to file a complaint that a law, regulation, or permit has been violated. There are also related provisions in some jurisdictions for a citizen to bring a statutory claim for damages where government or private action has, or will imminently, cause harm to the environment.

Private prosecutions

Another type of legal remedy to consider pursuing is private prosecution. Under Canadian law, citizens are entitled in some circumstances to stand in the shoes of the Attorney General and "privately prosecute" offences. As a result, where citizens have evidence on reasonable grounds to believe that another party has violated a law, regulation, or permit, they are entitled to swear an *information* setting out the facts of the violation before a court official (usually a Justice of the Peace).[12] If the Justice of the Peace concludes that the information discloses reasonable grounds to conclude that an offence has been committed, he or she is required to authorize a private prosecution to proceed.

Provincial policy with respect to private prosecutions varies. In some provinces, notably British Columbia, the Attorney General will intervene at this stage and will usually terminate (stay) the case. In other provinces (notably Ontario), however, such private prosecutions have been allowed to proceed and convictions have been secured. To assess the prospects of pursuing a private prosecution in your jurisdiction, it will be necessary to consult local legal counsel.

Judicial review

A third type of remedy that is frequently pursued by citizens and citizens groups is judicial review.[13] All administrative decisions made by government officials are subject to review by the superior courts to determine whether the decision has been made in accordance with the legal powers of the decision-maker. On satisfying the court that it should be granted standing, a citizen or citizens group can use judicial review to challenge the legality of emission, development, and siting permits. To proceed with such an action, it will ordinarily be necessary for the applicant first to pursue any statutory appeals that may be available to challenge the decision.

What is at issue during judicial review is not the wisdom of the decision but rather whether the decision is lawful with respect to the legal powers vested in the decision-maker. Grounds for judicial review might include allegations that the decision was made without proper consultation with the public, or that it was made for an improper purpose or was based on legally irrelevant considerations.

We can no longer take clean air for granted. Citizens everywhere are realizing this and are taking action.

Claims in tort

A final legal avenue that should be considered is to commence a common law action in tort.[14] Claims in tort can be brought by a person (the plaintiff) who has suffered personal injury, or whose use or enjoyment of their property has been wrongfully interfered with, due to the wrongful actions of another party (the defendant). Where the rights of large group of plaintiffs have been wrongfully affected, it may be procedurally advantageous to pursue the claim as a "class action" or "representative proceeding," depending on the laws applicable in the jurisdiction where the harm has occurred. It is important to recognize that tort actions are complex, time consuming, and expensive. On the other hand, if such an action is successful, the plaintiff(s) will normally be entitled to recover a monetary award to compensate them for the damage suffered and/or a permanent injunction against the offending activity.

Conclusion

Someone once said that not only can ordinary citizens fundamentally alter the *status quo*, but that the *status quo* rarely changes for any other reason.

We can no longer take clean air for granted. Citizens everywhere are realizing this and are taking action. What "action" means to each of us will vary greatly. This chapter and the Appendix that follows, we hope, will help you to realistically and strategically identify actions you can take that may make a difference. The rest, as they say, is up to you.

Endnotes

1 See web-site: www.ec.gc.ca/pdb/npri.html

2 See web-site: www.ene.gov.on.ca/envision/env_reg/er/registry.htm

Also, the *Ontario Bill of Rights* requires that proposed policies, Acts, regulations, and instruments be posted on an Environmental Registry.

3 *EBR* Registry Number: IA8E0887

4 See, for example, a successful appeal for a fee waiver brought against the Ontario MOE with respect to non-compliance records for municipal and industrial dischargers: Information and Privacy Commissioner (Ontario) order # P-1557 (April 27, 1998) . See web-site: www.ipc.on.ca

5 Environment Canada, *Community Environmental Projects: From Needs Assessment to Evaluation*, 1995. To order a copy, contact an Environment Canada Regional Office or download a copy from Environment Canada's Green Lane on the Internet at: www.ec.gc.ca (click on "reference guides").

6 Gahlinger-Beaune, R., *Not For Profit, You Say! An Operations Manual for Canadian Non-profit Organizations* (Burnaby, BC: Open-Up Poste Production, 1992).

7 Environment Canada, Community Animation Program. See web-site: www.ec.gc.ca/fund_e.html

8 Ura, M., *Making the News: A Guide to Using the Media* (Vancouver, BC: West Coast Environmental Law Research Foundation, 1992).

9 See web-site: canada.gc.ca/gazette/gazette_e.html

10 Tollefson, C., "When the 'Public Interest' Loses: The Liability of Public Interest Litigants for Adverse Costs Awards" (1996) 30 UBC Law Review 309

11 "Greenpeace Settlement Clears the Air," SLDF Newsletter (March 1997). See website: www.sierralegal.org

12 For further information on how to prepare and proceed with a private prosecution, see Werring, J., *Sierra Legal Defence Fund Handbook – How to Gather and Present Evidence for the Purposes of the Laying of Charges in a Private Prosecution Guide to Private Prosecutions* (Vancouver, BC: SLDF, 1997). Web-site: www.sierralegal.org/reprts/hanbook.html

13 For further discussion see Tollefson, C., "Public Participation and Judicial Review," in Hughes, E., et. al. (eds.) *Environmental Law and Policy,* 2nd ed. (Toronto, Ont.: Emond Montgomery Publications, 1998).

14 For further discussion see Charles, W., and VanderZwaag, D., "Common Law and Environmental Protection: Legal Realities and Judicial Challenges," in Hughes, E., et. al. (eds.) *Environmental Law and Policy,* 2nd ed. (Toronto, Ont.: Edmond Montgomery Publications, 1998).

Appendix

The following is a listing of primarily electronic sources containing updated and/or filed information on subjects and issues addressed in this book. They are organized as follows:

A. Government links – General
B. Environmental registries
C. Statutes and regulations
D. Freedom of information
E. Elected officials
F. Non-governmental links
G. Books and articles

A. Government links — General

- Federal government sites
- Provincial government sites
- US and international sites

Federal government sites

Commissioner for Environment and Sustainable Development (Office of the Auditor General of Canada). The Commissioner assists parliamentarians in their oversight of the federal government's efforts to protect the environment and foster sustainable development by providing them with objective, independent analysis and recommendations. www.dfait-maeci.gc.ca/sustain/Links/related-e.asp

Department of Justice (statutes and regulations; Attorney General; programs and services; public consultations): canada2.justice.gc.ca/

Environment Canada ("The Green Lane") – publications, news, events, related links – Environmental Registry; National Pollutant Release Inventory; Canadian Environmental Assessment Agency; Environmental Law Enforcement, etc.: www.ec.gc.ca

– Direct access to National Pollutant Release Inventory is: www.ec.gc.ca/pdb/npri/index.html

Supreme Court of Canada (decisions of Canada's highest court) www.scc-csc.gc.ca

Provincial government sites

(Environmental education, stewardship, trends; environmental legislation, projects, initiatives, reports; resource management; pollution standards and monitoring; popular topics, etc.)

Alberta Environment www.gov.ab.ca/env/

British Columbia Ministry of Environment, Lands and Parks www.gov.bc.ca/elp/

– Local governments also have authority over air issues. See, for example, the Greater Vancouver Regional District's web-site at: www.gvrd.bc.ca

Manitoba Conservation www.gov.mb.ca/environ/index.html

New Brunswick Department of Environment www.gov.nb.ca/environm/index.htm

Newfoundland and Labrador Department of Environment and Labour www.gov.nf.ca/env/Labour/OHS/default.asp

Northwest Territories Department of Resources, Wildlife and Economic Development www.rwed.gov.nt.ca/

Nova Scotia Department of Environment www.gov.ns.ca/envi

Ontario Ministry of Environment www.ene.gov.on.ca

– More detailed information on the Ministry is available at: www.ene.gov.on.ca/envision/org/org-moee.htm

Prince Edward Island Fisheries, Aquaculture and Environment www.gov.pe.ca/fae/index.php3

Quebec Ministere de l' Environnement www.menv.gouv.qc.ca

Saskatchewan Environment and Resource Management www.serm.gov.sk.ca

Yukon Territory Department of Renewable Resources
206.12.26.168/environ/

US and international sites

California Air Resources Board www.arb.ca.gov

US **Environment Protection Agency** (environmental news; programs; topics, e.g., air, cleanup, economics and ecosystems, environmental management; human health; pollution prevention actions; research; publications; laws and regulations; law enforcement, etc.) www.epa.gov

– Also, EPA has developed a Green Communities Assistance Kit, a step-by-step guide for planning and implementing sustainability at the community level. www.epa.gov/region03/greenkit/index.html

Commission for Environmental Co-operation was created under the so-called "environmental side agreement" to NAFTA. The CEC conducts and publishes research on environmental issues in the North American context. It also provides project funding to NGOs, and is a vehicle through which citizens can bring complaints about the non-enforcement of environmental laws by member countries (US, Canada, and Mexico). www.cec.org/infobases/law/index.cfm?format=2&lan=english

United Nations Development Programme (Sustainable Human Development) is a UN program aimed at helping countries in achieving sustainable human development the following areas: poverty eradication; employment creation and sustainable livelihoods; the empowerment of women; and the protection and regeneration of the environment. www.undp.org

United Nations Environment Programme has a mission "to provide leadership and encourage partnership in caring for the environment by inspiring, informing and enabling nations and people to improve their quality of life without compromising that of future generations." www.unep.org

The Intergovernmental Panel on Climate Change is a body established by the World Meteorological Organization and the United Nations Environment Programme. IPCC assesses the scientific, technical, and socio-economic information relevant for the understanding of the risk of human-induced climate change. www.ipcc.ch

B. Environmental registries

- Environmental assessment
- Environmental approvals, permits, licencing procedures

Environmental assessment

Canadian Environmental Assessment Agency prescribes and conducts environmental assessment of all new major physical works and activities within federal jurisdiction: www.ceaa.gc.ca

– Federal Environmental Assessment Index is a registry of documents relating to ongoing assessments: www.ceaa.gc.ca/registry/registry_e.htm

Provincial environmental assessment offices

Alberta Environmental Assessment/Evaluation
www.gov.ab.ca/env/protenf/assessment/index.html

British Columbia Environmental Assessment Office
www.eao.gov.bc.ca

– The Environmental Assessment Registry is directly accessible: www.eao.gov.bc.ca/project/home.htm

Manitoba Environmental Assessment Process www.gov.mb.ca/environ/pages/proposls/procbull.html

New Brunswick Environmental Impact Assessment Program
www.gov.nb.ca/environm/sciences/eia/index.htm

Newfoundland and Labrador Environmental Assessment
www.gov.nf.ca/env/Env/environmental_assessment.asp

Northwest Territories Environmental Protection Service
www.gov.nt.ca/RWED/eps/index.htm

Nova Scotia Environmental Assessment
www.gov.ns.ca/envi/dept/ess/ea/index.htm

Ontario Environmental Assessment Agency
www.ene.gov.on.ca/envision/env_reg/er/registry.htm

– The Ontario Environmental Assessment and Appeal Board is an administrative tribunal with a mandate to hear appeals of the Agency's decisions on various environmental assessments: www.ert.gov.on.ca/

Prince Edward Island Environmental Assessment
www.gov.pe.ca/te/ep-info/index.php3

Quebec Environmental Assessment
www.menv.gouv.qc.ca/programmes/eval_env/index.htm

Saskatchewan Environmental Assessment
www.serm.gov.sk.ca/environment/assessment/

Yukon Territories Environmental Assessment
206.12.26.168/environ/assess.html

Environmental approvals, permits, licencing procedures

Alberta www.gov.ab.ca/env/protenf/approvals/factsheets/approv.html

British Columbia does not have a central environmental permit system. Instead, one has to contact regional authorities directly. Some are not online and must be contacted by phone; others are electronically accessible. See, for example, Skeena Region Air Quality at: www.env.gov.bc.ca/ske/skeair/permits/permits.html, or www.elp.gov.bc.ca/ske/epp/eppperm_A.html

Manitoba Conservation Licensing and Approval
www.gov.mb.ca/environ/prgareas/apprv.html

New Brunswick and its Clean Air Act require registration of future polluters and a regime of approvals and permits: www.gov.nb.ca/environm/infoair/caa_description.html and www.gov.nb.ca/environm/infoair/PublicPP.html

Newfoundland and Labrador www.gov.nf.ca/env/env/envassess/guide%5Fto%5Fea.asp

Northwest Territories www.gov.nt.ca/RWED/eps/environ.htm

Nova Scotia www.gov.ns.ca/bacs/acns/paal/ndxenv.htm

Ontario Ministry of Environment – Operation Division, Approvals Branch: www.ene.gov.on.ca/envision/org/op.htm

Prince Edward Island www.gov.pe.ca/fae/index.php3

Quebec www.menv.gouv.qc.ca/programmes/eval_env/index.htm

Saskatchewan www.serm.gov.sk.ca/environment/assessment/reviewprocess.php3

Yukon Territory renres.gov.yk.ca/environ/air.html

C. Statutes and regulations

Federal

Canada Gazette is where all recent federal legislation and regulations are published: canada.gc.ca/gazette/gazette_e.html

Bills (draft legislation) are accessible at: www.parl.gc.ca/english/ebus.html

Federal environmental policy and planning documents www.ec.gc.ca/policy_e. html

Provincial statutes and regulations

Alberta www.gov.ab.ca/qp/

British Columbia www.qp.gov.bc.ca/stat_reg

Manitoba www.gov.mb.ca/chc/statpub/pricelist/index.html

New Brunswick www.gov.nb.ca/justice/asrlste.htm

Newfoundland and Labrador www.gov.nf.ca/just/

Northwest Territories www.assembly.gov.nt.ca/Legislation/index.html

Nova Scotia www.gov.ns.ca/legi/legc/

Ontario 209.195.107.57

Prince Edward Island www.gov.pe.ca/law/index.php3

Quebec www.justice.gouv.qc.ca/anglais/lois-a.htm

Saskatchewan www.saskjustice.gov.sk.ca/

Yukon Territories legis.acjnet.org/Yukon/index_en.html

D. Freedom of information

Under these statutes, citizens are provided the right to access government documents subject to certain exceptions.

Federal Government: Access to Information Act

insight.mcmaster.ca/org/efc/pages/law/canada/access.html

Alberta Freedom of Information and Protection of Privacy Act www.gov.ab.ca/qp/ascii/acts/F18P5.TXT

British Columbia Freedom of Information and Protection of Privacy Act www.qp.gov.bc.ca/bcstats/96165_01.htm

Manitoba Freedom of Information and Protection of Privacy Act www.gov.mb.ca/chc/statpub/free/pdf/f175.pdf

New Brunswick N/A

Newfoundland and Labrador Freedom of Information Act www.gov.nf.ca/just/

Northwest Territories Access to Information and Protection of Privacy legis.acjnet.org//TNO/Loi/a_en.html

Nova Scotia Freedom of Information and Protection of Privacy Act www.gov.ns.ca/legi/legc/index.htm

Ontario Freedom of Information and Protection of Privacy Act and Municipal Freedom of Information and Protection of Privacy Act www.gov.on.ca/MBS/english/fip/act/act.html.

Prince Edward Island N/A

Quebec Loi sur l'acces aux documents des organismes publics et sur la protection des renseignements personnels www.cai.gouv.qc.ca/loi.htm

Saskatchewan Freedom of Information and Protection of Privacy Act www.saskjustice.gov.sk.ca/

Yukon Territories Public Government Act (Part III: Freedom of Information and Protection of Privacy): legis.acjnet.org/cgi-bin/folioisa.dll/e_stats.nfo/query=freedom+of+information/doc/{@59146}?

E. Elected officials

The following sites provide access to elected officials at the federal, provincial, and territorial levels:

Parliament of Canada www.parl.gc.ca

Alberta www.assembly.ab.ca

British Columbia www.gov.bc.ca/bcgov/cont/

Manitoba www.gov.mb.ca/leg_asmb/index.html

New Brunswick www.gov.nb.ca/legis/index.htm

Newfoundland and Labrador www.gov.nf.ca/house/

Northwest Territories www.assembly.gov.nt.ca/

Nova Scotia www.gov.ns.ca/legi/housedir/alphlist.htm

Ontario www.ontla.on.ca/

Prince Edward Island www.gov.pe.ca/leg/index.php3

Quebec www.assnat.qc.ca/eng/indexne4.html

Saskatchewan www.legassembly.sk.ca/

Yukon Territories www.gov.yk.ca/legassem.html

F. Non-governmental links

- Canada
- US and international

Canada

- Environmental networks
- Other non-governmental organizations (NGOs)

Environmental networks

Environmental networks are national or provincial associations of environmental NGOs. Their objectives and activities include: promoting networking on environmental issues; fostering public involvement in legislative and policy processes; staging environmental conferences or similar gatherings; disseminating information and analysis; and coordinating joint action.

Canadian Environmental Network (CEN) www.cen.web.net/

Alberta Environmental Network www.web.net/~aen/

British Columbia Environmental Network (BCEN) www.bcen.bc.ca/

First Nations Environmental Network www.fnen.org/

Manitoba Eco-Network www.web.net/~men/

New Brunswick Environmental Network www.web.net/~nben/

Newfoundland and Labrador Environmental Network Youth Caucus www3.nf.sympatico.ca/nlen/INDEX.HTM or

Newfoundland and Labrador Environmental Network (e-mail address) cbnlen@nfld.bnet

Nornet (Yukon e-mail address) nornet@polarcom.com

Nova Scotia Environmental Network (e-mail address) nsen@web.net

Ontario Environmental Network www.web.net/~oen/

Prince Edward Island Environmental Network (PEIEN) www.isn.net/~network/

Réseau Québécois Des Groupes Écologistes www.cam.org/~rqge/ (French only)

Saskatchewan Eco-Network www.econet.sk.ca/

Other non-governmental organizations (NGOs):

Access to Justice Network (ACJNet) provides access to legislation, organizations, publications, databases, and discussion forums on justice and legal issues. www.acjnet.org

AirCare is a program aimed at motor vehicle emissions inspection and maintenance. Their site has information on testing centers, on-road testing, clean air tips. There is also a reference library available. www.aircare.ca

Auto Free Ottawa is a grassroots group of healthy life activists that promotes an urban lifestyle with reduced use of automobiles. www.flora.org/afo/

Better Environmentally Sound Transportation (B.E.S.T.) is a BC-based organization that promotes the use of sustainable and appropriate forms of transportation. www.sustainability.com/best/

British Columbia Public Interest and Advocacy Centre (PIAC) is a non-profit law office representing organizations and individuals in public interest cases before courts, administrative tribunals, and various levels of government. www.bcpiac.com

Canadian Centre for Pollution Prevention (C2P2) promotes the adoption of pollution prevention strategies in the government and private sectors. c2p2.sarnia.com/

Canadian Environmental Defence Fund (CEDF) is a national, charitable, non-profit organization founded to help citizens gain access to environmental justice. www.web.net/~cedf/

Canadian Environmental Law Association (CELA) CELA's main activities include: provision of equitable access to justice; advocacy for comprehensive laws, standards, and policies that will provide and protect environmental quality in Ontario and throughout Canada; facilitation of public participation in environmental decision-making; and provision of information, research, advice, and educational materials. www.web.net/cela

Canadian Institute for Environmental Law and Policy (CIELAP) is an independent, charitable, not-for-profit environmental law and policy research and education organization. www.cielap.org

Canadian Urban Transit Association (CUTA) is a professional association of urban transit advocates and supporters. CUTA has been a strong supporter of the idea of employer provided transit benefits and their tax exemption. www.cutaactu.on.ca

Clean Air Strategic Alliance (CASA) is an incorporated entity responsible for strategic management of air quality in Alberta. www.casahome.org/

Clean Nova Scotia is an educational organization that works to deliver environmental programs to Nova Scotia communities and empower Nova Scotians to promote environmental sustainability. www.clean.ns.ca/

Citizen's Environment Alliance of Southwestern Ontario (CEA) is a non-profit, grassroots, binational, education and research organization committed to an ecosystem approach to environmental planning and management. www.mnsi.net/~cea/index1.html

Conservation Council of New Brunswick primarily acts as citizen watchdog, safeguarding common wealth such as land, air, and water. Through research and education they develop and promote solutions to pollution and resource destruction. www.web.net/~ccnb/

David Suzuki Foundation (DSF) is a federally registered charity, which explores human impacts on the environment and looks for solutions that balance social, economic, and ecological needs. Its activities encompass: studying the causes of, and alternatives to, environmental threats; informing the public and decision makers about sustainable solutions; participating in projects and initiatives that serve as models for an ecologically balanced future. DSF is active in environmental research, education, and advocacy. www.davidsuzuki.org

Ecology Action Centre is one of the oldest NGOs in the Maritimes. Other than usual activities directed at the protection of the environment in its broadest sense, they are currently focussing on marine, wilderness, transportation, and environment/development issues. www.chebucto.ns.ca/Environment/EAC/index.html

Environmental Law Centre (Alberta) Society provides information and referral services (library, seminars, etc.) for individuals, NGOs, schools and universities, government, industry, and the media. Its lawyers are intensively involved with government consultation processes, task forces, round tables, and committees. The Centre also conducts research on a variety of environmental law and policy topics – including air pollution – and is an advocate for law reform. www.elc.ab.ca/

Friends of the Earth Canada is committed to working towards a sustainable global future through: research, advocacy, education, and cooperation (outreach to individuals, educators, governments, and other groups around the world). Friends of the Earth International has sixty-four sister organizations around the world. www.foecanada.org

Greenest City is a community-based NGO committed to reducing pollution, regenerating urban life, and promoting social equity. Their projects and campaigns embrace community diversity and engage people in finding locally appropriate solutions to global environmental problems. www.greenestcity.com/indexau.html

Greenpeace Canada is an independent, not-for-profit, campaigning organization that uses nonviolent, creative confrontation to expose global environmental problems. www.greenpeacecanada.org

Institute for Media, Policy and Civil Society (IMPACS) is an NGO based in Vancouver committed to the expansion and protection of democracy and the strengthening of civil society. Its goal is to help build strong communities by providing communications training and education to Canadian non-profit organizations, and by supporting free, open, and accountable media internationally. www.impacs.org

Nova Scotia Environment & Development Coalition works for sustainable communities worldwide. Through research, education, and advocacy, they highlight the connections among ecology, community economics, and social justice issues, and between the north and south. www.chebucto.ns.ca/Environment/NSEDC/index.html

Pembina Institute for Appropriate Development is an independent, citizen-based think tank, an activist public interest organization, and a consulting group with a reputation for technically reliable and innovative results. The Institute's members provide a multidisciplinary expertise in three major areas: energy and the environment, environmental economics, and sustainable resource management. www.pembina.org/

Pollution Probe is one of Canada's senior environmental organizations. Its main areas of activity are research, education, and offering of practical solutions to practical environmental problems. PP is focused on the reduction of smog-causing emissions, elimination of toxic chemicals in the environment, global climate change, and protection of human health. www.pollutionprobe.org/

Sierra Legal Defence Fund (SLDF) is Canada's largest public interest environmental law firm. It provides free expert legal advice and legal representation to environmentally concerned citizens and conservationist groups across Canada. Its long term goals include: advancement of strategic litigation (establishment of crucial legal precedents); coordination and cooperation of environmental interests and groups; development of effective legislation (e.g., endangered species protection, forest practices, and anti-SLAPP laws); and publication of reports and handbooks on environmental-legal issues. www.sierralegal.org

Sierra Club British Columbia is an NGO focused on exploration of the totality of environment (physical and spiritual) and restoration and preservation of the quality of human and natural environment in interactive harmony. www. sierraclub.ca/bc

Sierra Club of Canada is dedicated to exploring, enjoying, and protecting the wild places of the earth and to the practice of responsible use of the earth's ecosystems and resources. www.sierraclub.ca

Sustainable Development on Campus is a permanent conference of university leaders throughout the country dedicated to promoting environment protection within campuses, by students and faculties alike. iisd1.iisd.ca/educate/

Toxics Watch Society of Alberta is focused on environmental and public health issues related to toxic substances and pollution. Through a mixture of advocacy and community-based projects,

Toxics Watch strives to improve the quality of our air, water, and life. www.freenet.edmonton.ab.ca/compost/home.html

20/20 Vision BC is an NGO member of Earth Action, an international network of environmental and peace organizations that mobilizes ordinary citizens in building a safe, healthy future from an informed point of view. It is focused on small-scale actions, performed in goodwill, aimed at public decision makers. www.2020vision.bc.ca/

West Coast Environmental Law (WCEL) is a non-profit organization that is providing free legal advice, advocacy, research, and law reform services. Through its Environmental Dispute Resolution Fund, WCELhas donated over $2 million to hundreds of citizens groups across BC to help address environmental problems at the communty level. www.wcel.org

Corporate monitoring organizations

Micromedia, Ltd www.d-net.com/min/min40137.htm

Infomart Dialog Limited www.infomart.ca/

Ethics Scan Canada www.ethicscan.on.ca/

US and international

Amazing Environmental Organization Web Directory! is "the world's biggest environment search engine." www.webdirectory.com/

Center for Environmental Law and Policy is a non-profit membership organization dedicated to clean flowing waters for Washington. www.celp.org

Earthjustice Legal Defense Club (formerly the Sierra Club Legal Defense Fund) is "the law firm for the Environment." The Fund goes to court to safeguard public lands, national forests, parks, and wilderness areas; to reduce air and water pollution; to prevent toxic contamination; to preserve endangered species and wildlife habitat; and to achieve environmental justice. www.earthjustice.org

Environmental Media Services is an NGO that provides journalists with current information on environmental issues. EMS also provides media training to environmentalists as well as strategic communications support. www.ems.org

Natural Resources Defense Council is a non-profit organization with a primary objective of developing alternative ways of sustainability and general protection of natural resources. www.nrdc.org

The Rocky Mountain Institute helps individuals and the private sector to identify and implement new solutions to old problems. The emphasis is on the economics-environment relationship. www.rmi.org

Environmental Law Around the World (E-LAW) provides legal and scientific support to public interest environmental law advocates worldwide. www.igc.org/igc/issues/el/

Environmental Laws and Treaties is a comprehensive collection of international treaties and environmental laws from other countries compiled by Pace University Law School in New York. joshua.law.pace.edu/env/environ.html

World Resource Institute's primary goals are: to reverse the degradation of ecosystems; to halt the changes to the earth's climate caused by human activity; to catalyze the adoption of policies and practices that expand prosperity while reducing the use of materials and generation of wastes; and to fight for people's access to information and decisions regarding natural resources and environment. www.wri.org

G. Books and articles

Alternatives is a quarterly publication of the Faculty of Environmental Studies at the University of Waterloo, Ontario. E-mail address: alternat@watserv1.unwaterloo.ca

Atkinson, Michael M., *Governing Canada: Institutions and Public Policy* (Toronto, Ont.: Harcourt Brace Jovanovich Canada Inc., 1993)

Bates, David V., *A Citizen's Guide to Air Pollution* (Montreal, Que.: McGill-Queen's U.P., 1972)

Bates, David V., *Environmental Health Risks and Public Policy Decision Making in Free Societies* (Vancouver, BC: UBC Press, 1994)

BC Environmental Report is a quarterly published by the BC Environmental Network. E-mail address: info@bcen.bc.ca

Boardman, J. (ed) *Canadian Environmental Policy: Ecosystems, Politics, and Process* (Oxford, Eng.: Oxford University Press, 1992)

Durning, A., *The Car and the City* (Seattle: Northwest Environment Watch, 1996)

Estrin, D., and Swaigen, J., *Environment on Trial: A Guide to Ontario Environmental Law and Policy*, 3rd ed. (Toronto, Ont.: Emond Montgomery Publications, 1993)

Gahlinger-Beaune, R., *Not for Profit, You Say! An Operations Manual for Canadian Non-profit Organizations* (Burnaby, BC: Open-Up Poste Production, 1992)

Garrod, S., and Valiante, M., *The Regulation of Toxic and Oxidant Air Pollution in North America* (CCH Canadian Limited, 1986)

Harrison, K., *Passing the Buck: Federalism and Canadian Environmental Policy* (Vancouver, BC: UBC Press, 1996)

Hawken, Paul, *The Ecology of Commerce: A Declaration of Sustainability* (New York, NY: Harper Collins Publishers 1993)

Hilborn, J., and Still, M., *Canadian Perspectives on Air Pollution* (Ottawa, Ont.: Environment Canada, 1990)

Hogg, P., *Constitutional Law in Canada*, 2nd edition (Toronto, Ont.: Carswell, 1985). Rev. 1999

Hughes, Elaine, Lucas, Alastair, and Tilleman, William, *Environmental Law and Policy* (Toronto, Ont.: Emond Montgomery Publications Ltd., 1998)

Kindred, Hugh M., et al, *International Law Chiefly as Interpreted and Applied in Canada* (Toronto, Ont.: Emond Montgomery Publications Ltd., 1993)

Murphy, T., and Briggs-Erickson, C., *Environmental Guide to the Internet*, 4th ed. (Rockville, MD: Government Institutes Division, ABS Group Inc., 1998)

Rolfe, C., *Turning Down the Heat: Emissions Trading and Canadian Implementation of the Kyoto Protocol* (Vancouver, BC: West Coast Environmental Law Research Foundation, 1998)

Sandborn, C. Andrews, W., and Wylynko, B., *Preventing Toxic Pollution: Toward a British Columbia Strategy* (A Report to the BC Hazardous Waste Management Corporation) (Vancouver, BC: West Coast Environmental Law Research Foundation, 1991)

Scientific American (containing a wealth of accurate and current data on chemical pollutants, acid rain effects, measurements of pollutants worldwide, etc.): www.sciam.com

Transportation and Energy: Strategies for a Sustainable Transportation System (Berkeley, Cal.: American Council for Energy-Efficient Economy, 1995)

Ura, M., *Making the News: A Guide to Using the Media* (Vancouver, BC: West Coast Environmental Law Research Foundation, 1992)

World Commission on Environment and Development. *Our Common Future* (Oxford, Eng.: Oxford University Press, 1987)